CONTENTS

SPACE SHUTTLE SAFETY

WEDNESDAY, OCTOBER 1, 1997

U.S. HOUSE OF REPRESENTATIVES,
COMMITTEE ON SCIENCE,
SUBCOMMITTEE ON SPACE AND AERONAUTICS,
Washington, DC.

The Subcommittee met, pursuant to notice, at 10:05 a.m. in room 2318, Rayburn House Office Building, Hon. Dana Rohrabacher, Chairman of the Subcommittee, presiding.

Chairman ROHRABACHER. The Subcommittee on Space and Aeronautics is called to order.

In recent weeks we have all seen or heard several current or former NASA officials quoted in the media as saying that the risk of leaving David Wolf on board the Mir Space Station was less than the danger than he faced in riding up to Mir on the Shuttle. Whether or not that is true, it certainly reminds us of the importance of preserving and enhancing safety in the Shuttle program.

America's civilian space program began 39 years ago today. That is when NASA was created by Congress and President Eisenhower. And in that time, we have learned and relearned that you cannot take safety for granted or it will come back and bite you and hurt you. From Virgil Grissom to Christie McAuliffe, America's space pioneers have paid the ultimate tuition and we should have learned the lesson and we should heed that lesson.

Of course, no one can expect human space flight to be perfectly safe. Any time that human beings venture into a frontier, they take risks. The question is whether we are taking these risks responsibly with our eyes open.

Since Challenger, the U.S. space community has worked diligently to restore and then to preserve the safety of the Shuttle program. In recent years with guidance from the Craft Commission and a detailed analysis by SAIC and NASA, NASA and its contractors have been able to decrease reported anomalies while significantly streamlining and saving taxpayers dollars at the same time.

The Shuttle team deserves our praise, but so too, our continued vigilance. In particular, we must not allow other influences to distract our attention from safety. While it may travel through a near vacuum in low-Earth orbit, the Space Shuttle does not live on a policy vacuum at NASA. There are many pressures in the Shuttle program: to save money, to support as much scientific research as possible, to serve foreign policy needs, to launch the International Space Station on schedule, even to pursue new capabilities for future human space missions. Each and all of these pressures could have a significant impact on NASA and NASA's contractors efforts to maintain and enhance Shuttle safety.

And that is where this Subcommittee comes into the picture. It is not, and it can't be, our job to determine or guarantee safety of the Shuttle, but it is our job to oversee NASA, its contractors and its safety advisors to ensure that they are preserving Shuttle safety in the face of the aforementioned pressures.

The specific reason for this hearing is to provide our witnesses with the opportunity to report on the Shuttle and the Shuttle safety in context of repeated transfers of funds from the Shuttle program to the International Space Station. This spring NASA took $190 million out of the Fiscal Year 1997 Shuttle budget—and that was over the objections of this Committee, I might add—and this was done to pay for Russian non-performance of the International Space Station. And just last week, the Appropriations Conference Committee on VAHUD, acting at NASA's request, cut another $50 million from the Space Shuttle Program for Fiscal Year 1998 and gave it to the International Space Station. Since most of these funds were going to be spent on upgrades which would improve the Shuttle's safety and reliability, it seems self-evident that such cuts will have some impact on safety sooner or later. The only question seems to be, "How many more times can the cookie jar be raided before we get punished?"

NASA has told us that these transfers don't impact safety because they came out of funds carried over from previous fiscal years. In which case, the question is why NASA had such large carryovers to begin with? It seems at least a bit suspicious that on one hand NASA tells us that they need the entire Shuttle budget to preserve safety, but then extra money somehow appears exactly when the space Station needs it.

I think that everyone here knows that I want America to move as quickly as possible beyond the current launch vehicles—including the venerable Shuttle, I might add—and then into a new era of commercially developed fully-reusable space transportation vehicles. But until those are here, the Shuttle is how the U.S. government launches people and special payloads into space. So I have repeatedly supported NASA's full request for the Shuttle including near-term upgrades to improve safety and performance.

Either the Shuttle program needs its full budget to preserve safety or it doesn't. If the Shuttle managers are going to become loan officers for other NASA projects, then the bank's Board of Directors, meaning this Subcommittee, is going to consider investing those savings in new technologies to accelerate our progress towards cheaper and more reliable ways of getting into space rather than continuing to bail out continued Russian defaults or cost overruns for the International Space Station.

I would now like to recognize the Ranking Minority Member, Mr. Cramer of Alabama, to make his opening statement.

Mr. CRAMER. Thank you, Mr. Chairman. And I join with you, the Chairman of the Subcommittee, and the Chairman of the full Committee, in welcoming also our witnesses to this very important topic again for this Committee.

It's no exaggeration to say that the Space Shuttle has been the work horse of the Nation's human space flight program. Since its return to flight status following the Challenger tragedy, the Shuttle has achieved nearly a decade of safe and successful flights. It is a record that NASA can certainly be proud of, the contractors can certainly be proud of, and I'm proud of the participation of the Marshall Space Flight Center in this effort down in my home district of Alabama.

We all recognize that safety is something that requires constant vigilance, constant attention, especially during times of organizational change and budgetary stress. And we've certainly seen a little bit of that at NASA.

Today marks the first anniversary of the space flight operations contract that consolidated the Shuttle orbiter operations under a signal prime contractor, United Space Alliance. I'm looking forward to perhaps hearing a status report on that performance and whether anything else needs to be done to ensure continued safe operations of the Shuttle.

As the Chairman of the Subcommittee pointed out, I too am concerned about the Shuttle upgrades, what requirements they are intended to address about the transfer of funds from the Shuttle program and what impact that might have had or might have on the upgrades. So if you could consider that as you offer your testimony, that would be very helpful.

GAO has identified some potential issues related to the use of those carry-over funds to pay for some of those upgrades, and this Committee needs to hear from the GAO about its concerns as well. So we'll be listening attentively for that.

I thank the witnesses for their testimony in advance, and I look forward to an active hearing. Maybe not as many members as we would have hoped because we had a very late night last night. But I am sure that it will increase as the morning drags on.

Thank you, Mr. Chairman

Chairman ROHRABACHER. Thank you, Mr. Cramer. And I would like to congratulate Mr. Cramer for being so chipper this morning when the rest of us are dragging, but that does not include Mr. Sensenbrenner who is always ready.

Mr. Sensenbrenner, do you have an opening statement?

Chairman SENSENBRENNER. Thank you very much.

I would just like to make a couple of comments and pose some questions because I can't stay around for all of this hearing.

I'm very concerned that NASA is changing its tune relative to how much money is necessary to operate the Shuttle and how much money is necessary to ensure Shuttle safety.

On March 13, Mr. Johnstone, who is one of our witnesses, in his capacity as the Chairman of the Aerospace Safety Advisory Panel, testified before this Subcommittee, "Our view is that further reductions in funding could derail an otherwise well planned SFOC implementation. This could wreak hardships on the workforce and foster an overconfident or work around attitude, either of which could have serious safety implications. While safety is a popular by-word in the Space Shuttle Program, dollar constraints may limit the ability to perform tasks necessary to minimize risks."

Since Mr. Johnstone made that statement in March, NASA has looted the Shuttle account for $190 million during this fiscal year and is proposing to the Appropriations Conference Committee another $50 million for the next fiscal year. And I'm wondering which NASA we are to believe, the testimony that we're getting today or the testimony on the 13th of March.

Now, Mr. Trafton, for numerous years NASA witnesses including Administrator Golden have come before this Subcommittee to argue against cuts to the Shuttle's program telling us in no uncertain terms that it cannot endure further reductions without jeopardizing safety. At the same hearing in March, Steve Oswald, the Shuttle Program Manager, testified that any Shuttle savings should be spent on safety, "We should tend to try to upgrade the system and improve capability and safety in the Shuttle program." And on April 9, Mr. Trafton testified, "We do not intend to touch any 1998 dollars and beyond in the Shuttle program." And now we've got another $50 million hit.

And one of the things, gentleman, that this Subcommittee has done is defended the Shuttle program against raids from those in the Congress who would like to spend the money elsewhere either within or without NASA. What you have done by saying one thing in March, that we take to the Floor in the authorization bill and another thing today is destroy the credibility of the people who have defended NASA programs the strongest in the Congress of the United States. And I don't like it, and I want it to stop, and I hope that at today's hearing it will stop.

Thank you.

Mr. ROEMER. Mr. Chairman?

Chairman ROHRABACHER. Yes, Mr. Roemer?

Mr. ROEMER. May I ask unanimous consent to make just a 1-minute opening statement?

Chairman ROHRABACHER. I tell you what, we would be very happy to allow you and I understand Mr. Weldon, for short opening statements. So you may proceed.

Mr. ROEMER. Thank you, Mr. Chairman. I appreciate your patience and your willingness to let me talk for just a brief amount of time here.

I certainly concur that we have some worries about transfer authority here at NASA moving money from the Shuttle safety program to the Russian program. I have said time and time again to Mr. Goldin and to Mr. Culbertson and to others that we are increasingly worried about raiding accounts and having a space Station account gobble up other funds, kind of cannibalizing the NASA funds and risking the safety of our astronauts and the long-term safety of the NASA program.

We're making such great strides forward right now at NASA with the Rover, the Sojourner, gathering all that data and information. I hope we have a hearing on that to highlight NASA's great accomplishments there. But we seriously risk public support and the long-term efforts to rebuild confidence and excitement and the glory days of NASA if we continue to raid these accounts, other worthwhile accounts, to either transfer to the Russians because they are falling increasingly further behind. They are not coming up with the money to pay their Russian Space Agency. And now we read in Space News that the space Station appears to be $400-500 million over budget. Where is that going to come from? Is that going to come from Mission to Planet Earth? Is that going to come from Shuttle safety? Is that going to come from the development of our next more affordable and efficient rocket access into space to replace the Shuttle? I am increasingly worried that this lackadaisical transfer authority that we seem to give NASA to just transfer money at will from one account to another is going to erode Congressional confidence in NASA and public confidence in NASA. And I am very worried about the $400-500 million cost overrun. It just appears that NASA can say, "We seem to win space Station by enough votes, well we can be pretty cavalier about that program and spend money however we want, transfer money wherever we want and move forward to raid other very, very important and worthwhile NASA accounts."

And with that, Mr. Chairman, I conclude my brief opening remarks.

Chairman ROHRABACHER. The gentleman's concern is not misplaced either in terms of the practicality or philosophically. Because philosophically the people of this country did not invest the authority to determine where public funds would be spent with unelected officials at NASA. And when funds are transferred basically what we're seeing is the usurpation of Congressional authority, and it is very serious.

Mr. Weldon?

Mr. WELDON of Florida. Thank you, Mr. Chairman, and I want to in particular thank you for calling this hearing today. Your commitment to our Nation's space flight program is deeply valued, and I want to welcome our witnesses as well, particularly Mr. Trafton and Jim Adamson. It is good to see both of you here again.

I want to say at the outset how proud I am of the men and women who work at Kennedy Space Center in my district. The civil service and contractor teams there are the best in the world—that is being objective—and particularly at processing space craft and launching space craft. And my thanks go out to them and the dedicated teams that serve our Nation's space effort. They can be certain that I will always stand behind them and their work here on Capitol Hill.

I also want to specifically applaud the employees of United Space Alliance. They have worked to become more efficient at what they do, but at the same time have compiled an amazing track record of success in day-to-day operations of the Space Shuttle. I continue to wish them luck and will provide them whatever support they need. I congratulate you I believe on your 1-year anniversary on taking over the Shuttle operations which I believe is today.

Mr. ADAMSON. Thank you, sir.

Mr. WELDON of Florida. One of the things I hope to explore today is how these new efficiencies might enable the USA to pursue commercial opportunities on the Shuttle. I know that USA is today delivering a report to NASA discussing strategies to expand commercial and private use of the Shuttle, and I look forward to learning more about that, and I also think that would be an excellent topic for a hearing at some point, perhaps next spring.

So with that I conclude, and I yield back to the Chairman.

Chairman ROHRABACHER. I appreciate the statements from our colleagues. And without objection the written statements of other members will be made part of the written record. And with no objections it is so ordered.

[The prepared statements of Mr. Calvert, Ms. Jackson Lee and Mr. Lampson follow:]

Statement by Mr. Calvert
House Science Subcommittee on Space and Aeronautics

Thank you, Mr. Chairman.

In brief, I would like to express my sincere appreciation for holding this timely hearing.

I would also like to welcome our witnesses and thank them for taking some of their valuable time to be with us this morning.

Not too long ago, the full committee held an important meeting regarding the safety of the Mir Space Station.

Although many Members of Congress, including myself, expressed strong support for our nation's space program, many were still concerned about NASA's willingness to take on unnecessary risks with regards to this joint program.

Many felt that NASA was compromising its safety procedures for the sake of cooperating with the Russians.

With regards to the Space Shuttle program, I am satisfied that NASA is pursuing the right policy with regards to safety.

Since the Challenger disaster, numerous careful steps have been undertaken to ensure successful launches, and for that, NASA should be commended.

However, I know concerns still exist on how the consolidation of the shuttle program under one single prime contractor has affected Shuttle safety.

I would like to know how streamlining, which I assumed would be one of the results of having a single prime contractor, improved safety.

What safety policies were changed since the consolidation?

And how is the morale of the staff since the transition?

Again, Mr. Chairman, I thank you for holding this timely hearing and I look forward to their testimonies.

SHEILA JACKSON LEE
18TH DISTRICT, TEXAS

COMMITTEES:
COMMITTEE ON THE JUDICIARY
SUBCOMMITTEE ON CRIME
SUBCOMMITTEE ON COMMERCIAL AND ADMINISTRATIVE LAW

COMMITTEE ON SCIENCE
SUBCOMMITTEE ON SPACE AND AERONAUTICS
SUBCOMMITTEE ON BASIC RESEARCH

Congress of the United States
House of Representatives
Washington, DC 20515

WASHINGTON OFFICE:
410 CANNON HOUSE OFFICE BUILDING
WASHINGTON, DC 20515
202-225-3816
DISTRICT OFFICE:
1919 SMITH STREET, SUITE 1180
THE GEORGE "MICKEY" LELAND FEDERAL BUILDING
HOUSTON, TX 77002
713-655-0050

Statement
Congresswoman Jackson Lee
House Committee on Science's
Subcommittee on Space and Aeronautics
Hearing on Space Shuttle Safety
October 1, 1997

Thank you Chairman Rohrabacher and Ranking Member Cramer for this opportunity to receive testimony from witnesses on the current status of the Space Shuttle Program.

I like both of you am committed to a strong viable domestic space program. Today, we are doing the work of the 21st Century as we prepare for the reality of men and women working and living in space.

Our nation's space shuttle program was undertaken by NASA during the 1970s. This goal was to develop a space transportation system which could be used and reused to the maximum extent possible.

When Columbia, the first of NASA's orbiter fleet, was delivered to the Kennedy Space Center in March of 1979, a new era of space flight began. Those winged space craft were a culmination of over thirty years of work and an expression of how confident we were as a nation in our presence in space.

Columbia heralded the beginning of the modern American Space Flight program when it lifted off Pad A in the Launch Complex 39 area of the Kennedy Space Center on April 12, 1981. It proved the ability of winged, reusable spaceships by successfully completing that mission -- STS-1.

Since that first flight the Space Shuttle has flown over 80 missions carrying 212 of the 352 humans ever to have flown in space and has flown over 250 million miles.

This week from Pad 39-A the Atlantis lifted off for the latest mission of the orbiter fleet STS-86.

In order for the shuttles to be reusable they must be able to undergo changes which will keep them up-to-date with our ever changing space transportation experience.

The first shuttle orbiter, Columbia also became the first shuttle to undergo the scheduled inspection and retrofit program on August 10, 1991. Rockwell, the prime shuttle contractor, made approximately 50 modifications including the addition of carbon brakes, a drag chute, and improved nose wheel steering, to name a few of the improvements.

OPENING STATEMENT OF CONGRESSMAN NICK LAMPSON
OCTOBER 1, 1997
HEARING ON SPACE SHUTTLE SAFETY

Thank you Mr. Chairman. I welcome the opportunity to listen to the testimony of our distinguished witnesses -- Mr. Wilbur Trafton, Mr. Allen Li, Mr. Paul Johnstone, and Mr. Jim Adamson -- on overall Space Shuttle program safety, including whether the reallocation of funds from the Shuttle program to the International Space Station program has had any effect on ongoing upgrade programs or to the schedule for future upgrades.

We have all been witness to the many successful shuttle missions since April 1981, and also, unfortunately, to the tragic Challenger accident in 1986. As a result of the accident, the Rogers Commission recommended strengthening the oversight role of NASA headquarters, eliminating the isolation of program managers from the engineers, increasing the flow of information between the Shuttle workforce and the various program managers, and properly staffing and supporting a more robust safety organization within the program. I am pleased to note that NASA has since fostered an environment more conducive to open communications among the workforce and moved astronauts into program management.

On a different, but related, note, this past April NASA raised the issue of reallocating $190 million from the Shuttle program to the International Space Station in order to offset costs to that program as a result of Russian Space Agency monetary delays. I am quite confident in NASA's decision to reallocate these funds, and look forward to a response from Mr. Trafton as to whether this reallocation has had any, or will have any, effect on ongoing upgrade programs or to the schedule for future upgrades.

Again, thank you Mr. Chairman for calling this hearing to order. I look forward to hearing our witnesses' testimony, and the opportunity to ask a number of questions.

Chairman ROHRABACHER. The House is in session today, so we'll probably have to break for a vote, and without objection, the Chair asks unanimous consent to break and declare a recess at any time.

Hearing no objections, it is so ordered.

With us today are four distinguished witnesses, and they will be addressing different aspects of the Space Shuttle Program. And before I introduce them, I would like to ask all of them to rise so they can be sworn in.

If you could raise your right hand. Do you solemnly swear or affirm that the testimony you will give before this Subcommittee will be the truth, the whole truth and nothing but the truth? You may be seated.

Thank you all for being with us today. My staff tells me that it was a challenge to schedule you all at the same time, and we certainly understand that each of you is a busy person, and we appreciate you coming here and sharing your expertise with us.

Without objection, your written statements will be included in the record. So I would ask each of you to please summarize your statements taking out the most important aspects of your testimony and limit your testimony perhaps to about 5 minutes if possible.

First, we have Mr. Wilbur Trafton, NASA's Associate Administrator for Human Space Flight. In that position he has policy and program oversight responsibilities for both the Space Shuttle and International Space Station program.

We will hear testimony from each of our witnesses and then we will proceed with questions from the Committee.

Mr. Trafton, you may proceed.

TESTIMONY OF WILBUR TRAFTON, ASSOCIATE ADMINISTRATOR FOR HUMAN SPACE FLIGHT, NASA

Mr. TRAFTON. Thank you, Mr. Chairman

I appreciate this opportunity to address your concerns regarding our transfer of $190 million in Fiscal Year 1997 from the Space Shuttle Program to the International Space Station Program and what impact if any that transfer could have on Space Shuttle safety.

The $190 million was generated primarily from prior year operational efficiencies and program restructuring. Beginning in the Fiscal Year 1993 time frame, we began our Shuttle program restructuring process which included content deletions or the rephasing of some of our development projects.

As savings were realized in either operations or development, they were transferred from the projects into reserve accounts which were used to mitigate the impact of the significant reductions in NOA during Fiscal Year 1994 and 1995.

In my written statement, I have detailed the when, why, and how, of the funding transfers, so I won't go into the details at this time.

However, I would like to take a minute to emphasize that Space Shuttle safety, the Agency's number one priority has not been jeopardized by the transfer of funds to the International Space Station. Over the past few years, our budget has been reduced, but the Shuttle program has been accomplishing more complex missions routinely. While our critical metrics continue to show that we are increasing our safety margins and working better and smarter for less. We are managing the Shuttle program well and meeting our commitments to the taxpayers.

Last Thursday we launched STS 86, the seventh flawless, on time docking flight to Mir. In fact, we have not scrubbed a launch in nearly 2 years, and of the last 32 launches only 2 were outside the first 5 minutes of the launch window. This is something that many thought impossible 4 years ago. I am very happy to report that the current flight continues to go smoothly, but almost all Shuttle flights are like that recently, and our metrics verify the trend with indicators such as inflight anomalies and problem reports improving by factors of two or three since Fiscal Year 1993.

I am also happy to report that the SFOC transition continues to go smoothly and that the ASAP Panel represented here by Mr. Johnstone has given us a clean bill of health.

In addition, while increasing safety and meeting the manifest, we have also been able to help the International Space Station Program. Flying the Shuttle safely and successfully, assembly of the ISS are not only the one and two priorities of the agency, but they are becoming absolutely symbiotic in nature.

When it became apparent that the Station required additional funds in Fiscal Year 1998, I asked all programs under my purview to assess what flexibility they had in the near term to help out. The Space Shuttle Program recognized that not only is the Station dependent on a safe Shuttle to perform assembly and servicing missions, the Shuttle depends on the Station program for the majority of its scheduled seven to eight flights per year starting in Fiscal Year 1999.

Any significant interruption in ISS assembly would drive the Shuttle well below the 5 to 6 year minimum rate recommended to maintain a safe, proficient team. Therefore while it is critical to maintain or improve Shuttle safety, it is also critical that the ISS receive adequate funding to meet its schedule. And the Shuttle will do whatever it can.

Transferring the $190 million to the Station in Fiscal Year 1997 was the right thing to do, and we planned to transfer additional funds in Fiscal Year 1998. We believe that these transfers are consistent with the Agency's priorities and are in the best interests of human space flight.

With that, sir, I'll be happy to answer your questions.

[The prepared statement of Mr. Trafton follows:]

Hold for Release Until
Presented by Witness

Statement of

Wilbur C. Trafton
Associate Administrator
for Space Flight
National Aeronautics and Space Administration

before the

Subcommittee on Space and Aeronautics

Committee on Science
United States House of Representatives

October 1, 1997

Mr. Chairman and Members of the Committee:

Thank you for this opportunity to come before you to address the Subcommittee regarding our cost reductions initiatives within the Space Shuttle program coupled with our transfer of $190M in FY 1997 to the International Space Station (ISS). I will also explain how that transfer of funding will not compromise safety.

I feel that I can speak for all of us involved in Human Space Flight activities, from those of us here in Washington – to the Program managers at our various field centers -- to the technician on the floor of the Vehicle Assembly Building in Florida, flying the Shuttle safely is our number one priority. We all accept the fact that there is a certain amount of risk involved with flying humans into space but it is our job—all of us—to mitigate that risk. The Shuttle team, NASA and contractors, is a responsible, highly motivated, well trained, and extremely dedicated group of individuals. Our main goal is to assure that each and every Shuttle mission is prepared for and conducted in the safest manner possible.

Funding Transfer to the ISS

Some have asked how the Shuttle program can afford to transfer $190M to the ISS without affecting Shuttle safety? I would like to take a few minutes to explain that this transfer has had absolutely no effect on the safety of this treasured national asset or the crew members that fly aboard the Shuttle.

We began Restructuring the Space Shuttle program in FY 1992 consistent with the Agency Strategic Plan of reducing operations costs and to accommodate future declining budgets. Since that time we have reduced our budget by 30% equating to almost $1.3B in savings per year against FY 1992 budget levels. Restructuring will continue in the future and the Space Flight Operations Contract (SFOC) awarded last October is a large part of that. In fact, as you remember in the FY 1997 Congressional Budget, about $400M of our future run-out savings was due to SFOC. We have continued to assume those savings in both our FY 1998 and 1999 requests. During this restructure, the Space Shuttle program has tried to mitigate perturbations as much as possible. The two principle factors in this accomplishment are: (1) funding stability provided by the Congress over the last two years and (2) cost effective management by the Program as shown in **Attachment # 1.**

In addition to the Restructuring efforts, a variety of unusual circumstances occurred in FY 1996 which forced additional conservatism in the carryover estimates provided to GAO in support of the July 1996 hearing. It was the program's assumption that there was no new money available, therefore, they would need to fund any additional requirements. Specifically, low carryover estimates were provided due to impacts related to:

- Reversal of the "freeze design" decision in February 1996
- SFOC negotiations
- Unknowns in on-going development programs.

While the estimating assumptions were headed one direction, other factors contributed to actual spending being driven lower in the opposite direction, including:

- Most significantly, the sixteen Continuing Resolutions (CR) which occurred in FY 1996. In response, we directed that work should be accomplished to first ensure safety and then meet the manifest, while everything else waited until the CR was lifted in March of '97.
- Use of reserves was constrained due to the CRs, but also because of the on-going SFOC negotiations.
- The Government furlough postponed spending and contractual actions.
- Efficiencies in program operations were continuing.
- Super Lightweight Tank and Multifunctional Electronic Display System were underruning their cost estimates due to difficulties, since resolved.
- Alternate Turbopump program experienced programmatic delays.

Before the transfer, we had a projected carryover into FY 1998 of $550 million, plus or minus $50M. As you know, the carryover level required varies from program to program, with the amount dependent on procurement type and the level of risk. On average, a carryover of about 8 weeks is considered reasonable. The Shuttle program carryover Metric is 15%, a figure which is consistent with the Agency-wide average. To that number, we added about $70 million. In the Shuttle program, we also had about $70M set aside to continue a number of program approved upgrades and a $50M margin for program uncertainties.

Regarding the generation of the $190M transferred to ISS, these funds came primarily from prior year efficiencies in operations and from development projects' content deletions or re-phasing. As savings occurred, funds were transferred from the projects into reserve accounts, to mitigate perturbations on the projects in case of future funding reductions.

Now, at the end of FY 1997 and after the transfer of the $190M, we will be carrying $399M into FY 1998, which is sufficient to fund all of our projected Shuttle program requirements until the FY1998 Appropriation is received.

Of the $399M, our planned uses are as follows:

- $175M-200M minimum required carryover into FY 1999 (17 days)
- 75M Phase II upgrades
- 20M Advanced Space Transportation Technology studies
- 100M Estimating margins/planned transfer

SPACE SHUTTLE SAFETY

Budget planning and formulation for the Shuttle program is always accomplished consistent with our established program priorities : **1) Fly safely; 2) Meet the manifest; 3) Improve Supportability and 4) Reduce costs.** Safety is our number one priority and we all work together to ensure that the processes used to ensure safety remain strong and effective.

Following the Challenger accident and guided by the findings of the Report of the Presidential Commission on the Space Shuttle Challenger Accident - the "Rogers Commission Report" - NASA conducted an extensive, across-the-board effort to return to safe, reliable space flight. Numerous hardware, software, and process safety improvements were incorporated and both risk identification and reduction programs were put in place. The Space Shuttle program was reorganized to ensure that all necessary information was available to managers at all levels, i.e., communication channels were greatly enhanced. The Program Compliance Assurance and Status System (PCASS) and the Problem Reporting and Corrective Action (PRACA) system were set up to implement the information flow required.

All program documentation was reviewed, and previous waivers and deviations were revoked. Reestablishment required detailed analyses by both contractor and civil service personnel at all levels of management. Failure Modes and Effects Analyses were performed for every item, resulting in a Critical Items List. Fault trees were generated to characterize each subsystem in all the propulsion elements of the Shuttle. Disposition of every item reviewed was made through the Program Requirements Control Board system at each level of the program, with final disposition made by the Space Shuttle Program Director chaired Board. A detailed systems analysis was also accomplished for flight software. A review of every line of code was performed beginning with the Flight System Software Requirements (FSSR's). The FSSR's were reviewed and rewritten to enhance safe operation. In addition, an Independent Verification and Validation organization was established to verify that all new flight software was built according to these requirements. Also, hardware and software upgrades are continuing to provide a safer and more reliable vehicle. These additional upgrades will be carried to completion and the other enhancements will be sustained.

Critical to the success of the program following the Challenger accident are the processes which have been implemented to ensure complete and open communication of safety related issues. The commit-to-flight process epitomizes these principles. Each project reviews every technical issue and anomaly which could have any effect on the upcoming flight. Any technical issue is subjected to independent assessment by the government workforce. Through a series of open reviews, all significant issues and open items are elevated to a Flight Readiness Review Board chaired by the Lead Center Director for the Space Shuttle. This review is conducted "face-to-face" with all projects and senior government and contractor management in attendance. Through this process, full and open discussions of all potential issues are conducted and the issues are thoroughly assessed by all levels of program management, engineering, and safety and missions assurance.

In addition to the formal open communication system, a system is in place to provide a mechanism for anonymous, voluntary and responsive reporting of concerns or risks. The NASA Safety Reporting System (NSRS) is available to both NASA and contractor employees and has served as an effective supplemental reporting channel to augment existing means. The organization that processes the NSRS inputs is a private contractor. Shuttle related ground and flight safety concerns reported through the NSRS are worked directly by senior safety and program personnel. NASA believes that this system remains an effective "safety valve" that contributes to the overall Shuttle safety program. Also, it should be noted that anyone, at anytime, can stop the launch of a Space Shuttle if they suspect an issue of flight safety. No Shuttle will be launched until all concerns are successfully dispositioned.

We have also put into place a number of proactive initiatives which are designed to enhance safety awareness and training. For example, each October there is a program-wide Safety Awareness Day at all four of the field centers involved with the Shuttle program. Each Center approaches the day's activities in their own way with a variety of speakers, films, contests, and educational forums that are all focused on safety.

Additionally, organizational safety homepages have been developed and are on line. The *Shuttle Program Bulletin,* a US Alliance newsletter and numerous safety brochures are available throughout the workplace. A very active Space Flight Awareness program, focused on safety, provides a mechanism to reward individuals and organizations for superior performance. The Astronaut Corps is instrumental in the success of Space Flight Awareness by sponsoring highly regarded Awards and a Launch Honoree program to reward employees for their demonstrated dedication to flight safety.

All of these communication channels have been and will be maintained or improved even as we continue the transition toward Shuttle privatization. They are also endorsed, supported by, and actively participated in by the astronaut corps, which in itself is a tremendous motivational tool.

<u>Audit, Surveillance and Insight</u>

While government workers do not intend to continue to provide day-to-day oversight of routine operations, we will maintain in-depth knowledge of the integrity and health of contractor processes and products to insure safety and mission success. We will maintain technical insight into contractor activities, gathering product or process data that provides adequate visibility into the integrity of the process. To provide this greater insight into contractor operations, the NASA Engineering and Quality Audit (NEQA) program, an improved approach to audit using engineering and quality personnel to verify stability of processes, has been implemented. The NEQA program uses a two-phase audit approach (contractor self-audit and NASA on-site audit), providing a method for evaluating processing systems and procedural interrelationships, the effects of management systems, and hardware controls. In addition to NEQA, in-plant surveillance will be utilized, including but not limited to selected high risk areas such as proof test results, systems tests, checkout, and validation process control, training, and configuration management.

All technical issues and any problems or anomalies (occurring either during ground processing or in flight), which are outside of our experience base and/or have not been previously addressed will be subjected to independent assessment by government personnel. All such issues and the corresponding independent assessments will be included in the Flight Readiness Review process.

Metrics

Metrics have been established to monitor parameters which indicate the health of the program and are indicative of safety related trends. These include in-flight anomalies, cannibalization error rates, and operational problem reports. These metrics and others will continue to be monitored and corrective action taken where problems are indicated. Thus far, in-flight anomalies have decreased from a per flight average in 1993 of 14.3 to a current average of 6.75--an improvement of over 50%. The same trend is true for monthly mishap frequency at Kennedy Space Center--a decrease of over 40% in three years. Other metrics are revealing similar trends, leading to the conclusion that we are doing things better but not at the expense of safety. We work closely with NASA Safety and Mission Assurance personnel at Kennedy and all other Shuttle facilities where hardware is built, processed or operated to assure that the Program meets their stringent requirements and is operating with established metrics.

Role of Safety and Mission Assurance (S&MA)

Currently NASA Safety and Mission Assurance (S&MA) conducts an independent commit-to-flight process which stresses assessment of all Shuttle processing activities and known issues and constraints. It consists of three incremental mission specific S&MA reviews prior to each Space Shuttle launch. Pre-launch Assessment Reviews, Flight Readiness Review tag-ups, and Launch Minus 2 Days tag-ups keep all organizational levels of S&MA apprised of status, progress, issues, concerns, and remaining open work to be accomplished before a mission is certified to fly. The S&MA commit-to-flight process has been refined over time and is robust and stable. It has been successfully executed since Return-to-Flight and it will continue to ensure that all critical path processes are verified for flight readiness.

NASA S&MA is transferring responsibility and accountability to the contractor S&MA for selected tasks currently performed by NASA and/or the Defense Contract Management Command (DCMC). NASA S&MA will assure that the transitions are conducted in a controlled, effective and reliable manner. NASA S&MA studies have shown that redundant Government Mandatory Inspection Points (GMIP's) dilute accountability and do not ensure achieving a successful process. The GMIP's will be reduced or phased out of stable processes over the next few years. The contractor will perform 100% of the critical inspections while the government S&MA and engineering organizations will conduct a robust surveillance program.

NASA S&MA will rely more on validating the processes the contractor uses to conduct in-plant work and less on inspecting the final product. NASA SMA is a key member of the aforementioned NEQA audit team. Also, DCMC and NASA resident office S&MA personnel will be co-located in work teams, work centers, and work areas as necessary to monitor processes, documentation, and hardware.

Independent Assessments

The Shuttle program has continued to increase safety margins. This has been confirmed by a number of independent assessments since 1993. Most recently, the Administration, together with the NASA Administrator, requested that the Aerospace Safety Advisory Panel (ASAP) perform an independent and in-depth review to verify the program's health. The study identified twenty-two recommendations in the areas of Maintenance of Independent Safety Oversight, Lead Center/Communications, Downsizing, Space Flight Operations Contract, and Manifest/Shuttle upgrades. As of August 18, 1997, only one recommendation remains open pending a follow-up visit to the Kennedy Space Center. The Panel found that the program was robust and recommended that we continue to be cautious and to encourage more open communications between management and workers on the line. We appreciate the thoroughness and professionalism with which the ASAP conducts their reviews and look forward to working with them in the future to assure and safe and reliable Space Shuttle program.

Conclusion

In closing, I would like to re-emphasize that Space Shuttle safety, the Agency's #1 priority, has not been jeopardized by the transfer of funds to the International Space Station program. The assembly of the ISS is the Agency's #2 priority and is absolutely dependent on the Space Shuttle for successful completion. Therefore, it is critical to maintain or increase Shuttle safety margins to better support ISS assembly, and we plan to do so. However, at this point in NASA's Human Space Flight endeavors, the Shuttle and Station programs have become symbiotic. The vast majority of the Shuttle's 7-8 flights per year after FY 1999 are devoted to ISS assembly and servicing. Any significant perturbation to Station would reduce Shuttle to well under the minimum rate of 5-6 flights per year recommended to maintain a safe and proficient workforce. Therefore, consistent with the Conference Agreement accompanying the FY 1998 VA-HUD-Independent Agencies Appropriations bill, we are prepared to transfer an additional $50M of Shuttle funding to ISS in FY 1998, consistent with the Agency priorities.

45-568 98 - 3

Chairman ROHRABACHER. We'll have questions for the whole panel when everyone is done.

We have next, Mr. Paul Johnstone who is the Chairman of NASA's Aerospace Safety Advisory Panel. And that panel provides NASA and us with an ongoing independent review of safety issues for programs like the Space Shuttle.

Mr. Johnstone, please proceed.

TESTIMONY OF PAUL JOHNSTONE, CHAIRMAN, THE AEROSPACE SAFETY ADVISORY PANEL, NASA

Mr. JOHNSTONE. Thank you, Mr. Chairman.

Distinguished members of the Subcommittee, you have my written testimony, therefore I will limit my remarks to a couple of major points.

Earlier this year it came to the panel's attention that a portion of the prior year's Space Shuttle appropriated funds was being applied to the International Space Station Program. Further the panel was asked to provide a viewpoint as to whether such a transfer would impact Space Shuttle safety. The panel responded in an April 14, 1997, letter to the NASA Administrator. A copy of that letter was provided to your staff.

Our basic message was and still is that the panel does not review budgetary manners and relies upon the affected programs to provide assessments of the consequences of funding changes on operations. The Space Shuttle Program has not informed the panel of any functional changes as a result of this uncosted carry over funds transfer. Our position with regard to all overall Space Shuttle safety has not been altered since my testimony to you on March 13, and the panels two most recent publications serve as background references to our position. They are the Annual Report of 1996 and the Review of Issues Associated with Safe Operations Management of the Space Shuttle Program, the so-called White House Study. Our next report will be published in February 1998.

The panel is aware of the potential implications of a funds transfer. For instance, if planned Space Shuttle safety upgrades were to be deferred or outright canceled, this could impact the ongoing efforts to reduce risk. At present we do not have any indication that this funds transfer would have such an effect. We will continue to watch the situation.

Your letter asked about current status of upgrade programs for the Space Shuttle. As you are undoubtedly aware, NASA has adopted a phased approach to this effort. The initial upgrades are to improve safety and provide additional performance for ISS assembly and utilization.

The panel continues to closely monitor these initiatives. For example, I along with two panel members attended the super lightweight tank design certification review yesterday at Michoud. Our computer hardware/software team was at the IV&V facility in Fairmont on Monday, and another team was out at Boeing Rocketdyne and Space Systems Division last week for briefings on the Space Shuttle main engine and orbiter upgrades program. The panel will continue to keep a close eye on the safety aspects of all upgrades that NASA proposes.

During our visits to NASA centers and contractors, briefings are requested that outline Space Shuttle upgrade plans. Follow-up fact finding is then done by individuals or subteams of the panel. NASA, United Space Alliance and other contractors have been very cooperative in meeting the panel's request for information.

The slippage of the ISS assembly schedule was the next topic you asked me to address. The revised schedule has provided an opportunity for some integrated testing of components on the ground prior to launch. We get the "ship and shoot" philosophy that had been going on.

The panel applauds this added step as it should greatly increase the quality of the hardware on orbit and lesson the tasks of the orbital assembly crews. As a result, risks should be reduced.

However, this additional testing at Kennedy Space Center has the potential to strain some Space Shuttle personnel resources. The panel will watch this situation in the coming months and report to NASA if we see any problems developing. Subteams are visiting KFC again next week.

The final topic you asked for the panel's view on was the issue of replacement parts and loss of vendor support.

NASA is in the best position to provide an overall assessment of this situation. The panel has taken a key interest in this issue. Our Annual Report for 1996 contained a finding which states, "obsolescence of components and systems on the Space Shuttle is an increasing problem threatening critical spares availability." Our recommendation was "alternative components must be developed and certified and where necessary systems must be redesigned to use available or adaptable units."

NASA has acknowledged that it will continue to coordinate obsolescence issues concerning hardware, special test equipment, vendor capability, environmental restrictions with the appropriate design centers. The program is designed to assure that potential problem areas are addressed in a timely manner. The panel will keep apprised of that and the resolution of ongoing concerns and will regularly document its findings.

In closing I would like to point out that in spite of all the changes that NASA is going through, from downsizing to restructuring its mode of overall program management, we continue to note that safety remains the number one tenet of the NASA and contractor Space Shuttle team.

That concludes my opening statement. I would be pleased to answer any questions when the time arises.

[The prepared statement of Mr. Johnstone follows:]

Hold for Release Until
Presented by Witness

Statement of

Mr. Paul M. Johnstone

Chairman, Aerospace Safety Advisory Panel

before the

Subcommittee on Space and Aeronautics

Committee on Science

United States House of Representatives

October 1, 1997

Hold for Release Until
Presented by Witness

Statement of

Mr. Paul M. Johnstone

Chairman, Aerospace Safety Advisory Panel

before the

Subcommittee on Space and Aeronautics

Committee on Science

United States House of Representatives

Mr. Chairman and Distinguished Members of the Subcommittee:

I am pleased to appear before you today to share the Aerospace Safety Advisory Panel's view regarding: the impact of a proposed transfer of funds from the Space Shuttle program to the International Space Station (ISS) program, the status of the Space Shuttle program safety and performance upgrades program, the effects of ISS schedule slippage on Space Shuttle launch requirements and the status of the Space Shuttle program with respect to spare parts and vendor support.

The Panel realizes that NASA faces significant funding challenges in building and operating the ISS. Along with this very ambitious undertaking, the continued safe operation of the Space Shuttle is an absolute necessity. There are, and undoubtedly will continue to be, tradeoffs among these two major programs as to how the precious dollars for human spaceflight should be allocated.

Earlier this year it came to the Panel's attention that a portion of prior year Space Shuttle appropriated funds, i.e., uncosted carryover, was to be applied to the ISS program. Further, the Panel was asked to provide a viewpoint as to whether such a transfer would impact Space Shuttle safety. By way of a letter addressed to the NASA Administrator, the Panel's position was stated. This April 14, 1997, correspondence has been provided to your staff.

The Panel does not review budgetary matters and relies upon the effected programs to provide assessments of the consequences of funding changes on operations, e.g., the impacts of a funds transfer from Space Shuttle to ISS. The Space Shuttle program has not informed the Panel of any functional changes as a result of this uncosted carryover funds transfer. I testified before your Subcommittee on Space and Aeronautics on

March 13, 1997, as to the Panel's view towards the overall safety of the Space Shuttle program. Our overall views have not been altered since then. The primary background references for this position are the Panel's two most recent publications: the *Annual Report for 1996* and the *Review of Issues Associated with Safe Operation and Management of the Space Shuttle Program.*

The Panel is aware of the potential implications of a funds transfer. For instance, if planned Space Shuttle safety upgrades were to be deferred or outright canceled, this could definitely impact the ongoing efforts to reduce risk. At present, we do not have any indication that this proposed funds transfer would have such an effect. We will, however, continue to watch this situation.

Over the years, the Panel has actively monitored the safety upgrades for the Space Shuttle. The overall program plan for the upgrades appears to be well conceived and is one that we believe will meet the need to support the ISS. In the aforementioned *Review of Issues Associated with Safe Operation and Management of the Space Shuttle Program,* the Panel does recommend, however, that "NASA should ensure that all enhancements and upgrades are fully tested and assessed prior to implementation. Funding profiles and schedule pressures should not be allowed to shorten any critical testing or validation processes." Two months ago, during a program status update the Panel was assured by the Space Shuttle Program Manager that no testing/validation "shortcuts" will be allowed. It is also important to note that some of the items classified as "upgrades" are also vitally needed solutions to problems of obsolescence and loss of vendor support as well as being improvements of the baseline system

A phased approach to Space Shuttle upgrades has been implemented. The so-called "Phase I" upgrades are safety and performance improvements to support the ISS assembly and utilization. Examples of these are the Space Shuttle Main Engine (SSME) large throat main combustion chamber and alternate turbopump upgrades. The Panel has closely monitored these SSME upgrades and will continue to do so to assure that safety certification criteria are met. The development of a new External Tank for the Space Shuttle, called the Super Light Weight Tank (SLWT), is another Phase I upgrade that the Panel has looked into. The new aluminum-lithium alloy used in the tank walls and domes has different fracture toughness characteristics and welding properties than that of the current aluminum tank. Numerous discussions between the Panel and the NASA and contractor experts involved in the tank design and testing have been conducted. The Panel believes that strict adherence to the quality control procedures and proof testing criteria will result in the safe utilization of the SLWT.

Phase II upgrades are those that NASA considers as high value, low impact, and incremental in nature. An example in this category includes the incorporation of the global positioning system (GPS) in the orbiters. NASA is on track to have each orbiter outfitted with GPS during its next scheduled maintenance down period. Another Phase II upgrade is the multi-function electronic display system (MEDS). This development and installation of a "glass cockpit" suite of displays for retrofit in the orbiters is on schedule, with the first planned flight on Atlantis in early 1999.

The study and definition for Phase III upgrades, those deemed as high value, major system upgrades with more extensive implementation impact and cost, have begun this year. Phase IV upgrades are those of high value and of such a major impact that modification to vehicle configuration would result. The prime example of this type would be the incorporation of the liquid fly back boosters many years in the future.

The Panel will continue to keep a close eye on the safety aspects of all upgrades that NASA proposes. During our visits to the NASA centers and prime contractors, briefings are requested that outline Space Shuttle upgrade plans. Follow-up fact-finding, as required, is then done by individuals or sub-teams of the Panel. NASA, United Space Alliance and other contractors have been very cooperative in meeting the Panel's requests for information. Of recent note are the detailed briefings on the SLWT certification plans. In order to assure the most complete understanding of SLWT flight readiness, the Panel members who have closely followed this particular program also attended the Design Certification Review that was held yesterday at the Michoud Assembly Facility.

The slippage of the ISS assembly schedule has provided an opportunity for some integrated testing of Space Station components on the ground before launch. We are in favor of this added step as it should greatly increase the quality of the hardware on orbit and lessen the tasks of the crews who must assemble the ISS. As a result, risk should be reduced. This additional testing at the Kennedy Space Center (KSC), however, has the potential to strain some Space Shuttle personnel resources. In particular, the Safety and Mission Assurance function at KSC may become taxed by the need to support the launch manifest while simultaneously "lending" personnel to assist in the integrated ISS testing. We will watch this situation carefully in the coming months and report back to NASA if we see any problems developing.

NASA is in the best position to address the details of the issue of Space Shuttle replacement parts and loss of vendor support that was mentioned in your invitation letter for this hearing. I would like to emphasize to this Committee that the Panel has taken a keen interest in this particular situation. In our *Annual Report for 1996*, a finding was cited as follows: "Obsolescence of components and systems on the Space Shuttle is an increasing problem threatening critical spares availability." Our accompanying recommendation stated: "Alternative components must be developed and certified, and, where necessary, systems must be redesigned to use available or adaptable units."

NASA has acknowledged that it will continue to coordinate obsolescence issues concerning hardware, special test equipment, vendor capability, and environmental restrictions with the appropriate design center. Additionally, NASA has assured the Panel that the Space Shuttle upgrade program is designed to assure that potential problem areas are addressed in a timely manner so as to not disrupt mission requirements. The Panel intends to keep apprised of the resolution of these ongoing concerns and will regularly document its findings.

NASA is undergoing a great deal of change from downsizing and the restructuring of its mode of overall program management. The Panel is pleased to note, however, that safety clearly remains the number one tenet of both NASA's dedicated workforce and its contractor teammates.

Mr. Paul M. Johnstone

Chairman, Aerospace Safety Advisory Panel

Mr. Paul Johnstone spent his career in engineering design, development, acquisition, and operation of commercial aircraft. After obtaining his B.S. degree in Aeronautical Engineering from the University of Notre Dame in 1946, he joined the Douglas Aircraft Company, where he was first an aircraft performance engineer and then a stability and control engineer engaged in aircraft design and certification. After 5 years as an Operations Engineer and later head of Technical Operations for Hawaiian Airlines, he joined Eastern Airlines as Manager of Economic and Performance Analysis in the Development Engineering Group. At Eastern, Mr. Johnstone rose to Vice President, Engineering, and later Senior Vice President, Operations Services. In those positions, he directed evaluations and was responsible for engineering, quality assurance, maintenance, inventory management and control, production planning and control and purchasing. In addition, he was corporate technical representative at all Eastern accident investigations. Mr. Johnstone is a Fellow of the American Institute of Aeronautics and Astronautics. Mr. Johnstone became a member of the Aerospace Safety Advisory Panel in September 1992, after joining the Panel as a consultant in 1991. He has been Chairman since March 1995.

Chairman ROHRABACHER. Thank you very much, Mr. Johnstone.

Third, we have Allen Li, an Associate Director for the U.S. General Accounting Office which has recently completed a study of NASA's Space Shuttle budget. Mr. Li has served as a source of information and research for this Subcommittee in the past and for actually other committees that I have chaired in the past as well, and his expertise and his diligence is well appreciated by this Chairman, and we appreciate him being with us today to share with us his findings.

Mr. Li?

TESTIMONY OF ALLEN LI, ASSOCIATE DIRECTOR, U.S. GENERAL ACCOUNTING OFFICE

Mr. LI. Thank you for those kind words, Mr. Chairman.

Mr. Chairman and members of the Subcommittee, I am pleased to be here today to discuss two aspects of NASA's Space Shuttle Program, upgrade activities and carryover balances. As requested, I will summarize my prepared statement.

I agree with Mr. Trafton with regard to the importance NASA places on Shuttle safety. The Space Shuttle is the Nation's only means for transporting humans to and from space. The Shuttle will be required well into the next century for space Station assembly and operation support and other human space flight missions. As such, that is why the Agency plans and implements Shuttle upgrades to improve safety and reliability.

But upgrades are also made to overcome component obsolescence, enhance Shuttle performance, and reduce operations cost. Upgrades will be necessary until the Shuttle is replaced.

Last year we testified before this Subcommittee on NASA's use of carryover balances, funds that are used in a subsequent year. Carryover balances consist of unobligated funds and uncosted obligations. Our position then, as it is today, was that while some carryover balance is appropriate programs, more indepth review is needed on the justifications for carryover balances in order to provide the context for assessing these balances.

I will make three points this morning. Point one, the $190 million funding transfer to the Space Station Program did not adversely impact current or supportability upgrade projects. Four current projects which account for about 91 percent of the total funding provided for upgrade activities in Fiscal Year 1997 could not have used the transferred funds. Even though they experienced technical problems and schedule slippages, the projects' financial reserves were enough to fund the problems experienced. An example of a current upgrade project is the alternate high-pressure turbo pump.

The funds transferred to the space Station were also not needed for supportability upgrades because these started more slowly than planned. Some of the supportability upgrades approved so far include the checkout and launch control system at Kennedy and a study on liquid flyback boosters.

My second point, the $190 million transfer to the space Station was available because of the large amount of carryover within the Shuttle program. Without the transfer, the Shuttle program would have had about a $600 million budget carryover at the end of Fiscal Year 1997. In May of 1997, Shuttle program officials indicated that they would like to have a carryover balance of $395 million at the end of Fiscal Year 1997 including an amount to cover upgrades.

NASA has used carryover to fund upgrades in the past and plans to continue doing so in the future. For example, NASA designated $70 million in Fiscal Year 1996 carryover funds to initiate the new supportability upgrade effort in Fiscal Year 1997. It plans to use another $95 million for this purpose in Fiscal Year 1998.

My third and final point: Depending on the upgrade selected, future costs could range from hundreds of millions to several billions of dollars. Which upgrades and who will foot the bill are open questions.

NASA is presently defining an upgrade program to keep the Shuttle a viable space transportation system at least through the planned mission life of the space Station, about 2013. Other planning assumptions included increasing the Shuttle's flight rate from about 8 to 15 times a year, use by DOD and commercial customers, and operating the Shuttle to 2030. In addition to annual funding requests for safety and performance upgrades and the continued use of carryover funds, NASA plans to use any savings that are generated within the Shuttle program to fund its upgrade activities.

However, the extent to which these sources will be available in the future is uncertain. So funding is the key issue. Funding needs for future Shuttle upgrades will be driven by a number of questions. These include: How long will the Shuttle be required? If the Shuttle is privatized, who should pay for the required upgrades? Can viable alternative be developed to support space operations and other human space flight requirements?

We believe that ongoing analysis such as the common space transportation technology development plan will serve to initiate the necessary debate for shaping the Nation's future transportation policy. The results of that debate may help provide answers to these questions.

Mr. Chairman, this concludes my summary.

[The prepared statement of Mr. Li follows:]

United States General Accounting Office

GAO

Testimony

Before the Subcommittee on Space And Aeronautics, Committee on Science, House of Representatives

For Release on Delivery
Expected at
10:00 a.m., EST
Wednesday,
October 1, 1997

SPACE SHUTTLE

Upgrade Activities and Carryover Balances

Statement of Allen Li, Associate Director, Defense Acquisitions Issues, National Security and International Affairs Division

GAO/T-NSIAD-98-21

Mr. Chairman and Members of the Subcommittee:

We are pleased to be here today to discuss two aspects of the National Aeronautics and Space Administration's (NASA) space shuttle program–upgrade activities and carryover balances.

In May 1997, the House Science Committee expressed concern about NASA's plans to take $190 million from the space shuttle program to help offset additional costs in the International Space Station program. According to NASA, the funds were available for transfer because of the high carryover balances projected to be available at the end of fiscal year 1997. The funding transfer was made, with a significant portion of the funds coming from the safety and performance upgrade portion of the shuttle program. Pursuant to the Committee's request, we agreed to review (1) the effect of the funding transfer on major shuttle upgrade projects and the status of those projects, (2) the role carryover balances had in the transfer and in funding upgrades, and (3) funding for future upgrades.

BACKGROUND

The space shuttle is the United States' only means for transporting humans to and from space and NASA considers safe operation of the shuttle as its top priority. The shuttle has flown 86 missions over the past 16 years. It will be required well into the next century for space station assembly and operations support and other human space flight missions. Upgrades will be a necessary part of the program until the space shuttle is replaced. NASA's strategy is to identify and prioritize upgrades to meet the program goals of (1) improving system safety, (2) supporting the flight manifest, (3) improving system supportability and reliability, and (4) reducing system operating costs.

NASA began a new supportability upgrade program in fiscal year 1997 to combat obsolescence in, and improve the reliability of, the space shuttle. This program involves updating and modernizing shuttle components, such as the checkout and launch control system at Kennedy Space Center. According to NASA, supportability upgrades are required to help ensure continued safe shuttle flight operations.

Future plans for the shuttle that could influence upgrade decisions are currently being formulated. At the direction of the Office of Management and Budget (OMB), NASA is developing a space transportation strategy. In support of that strategy, NASA is also developing a common space transportation technology development plan. Taken together, these plans are intended to guide decisions related to future investments in space transportation systems. The strategy is to be coordinated with both the Department of Defense and the commercial sector to ensure an integrated approach to U.S. space transportation in the 21st century. These documents are to accompany NASA's fiscal year 1999 budget submission to OMB.

NASA's budget for the space shuttle in fiscal year 1997 was $3.15 billion, including $636 million in the safety and performance upgrades line item. Space shuttle funding is available for obligation over a 2-year period. Funding that cannot be used in the year it is appropriated can be carried over to help cover program costs in the next fiscal year. Carryover includes unobligated funds as well as funds that have been obligated for goods and services but for which costs have not been incurred. The amount of carryover at the end of a fiscal year is determined by the total budgetary resources available less the costs incurred that year. Total budgetary resources consist of carryover from the prior year plus new budget authority.[1]

---

[1]For a discussion of carryover balances, see NASA Budget: Carryover Balances in Selected Programs (GAO/NSIAD-96-206, July 16, 1996) and (GAO/T-NSIAD-96-207, July 18, 1996).

SUMMARY

The $190 million funding transfer to the space station program did not adversely impact current or near-term upgrade projects. Four current projects, which account for about 91 percent of the total funding provided for upgrade activities in fiscal year 1997, could not have used the transferred funds. Even though they experienced technical problems and associated schedule slips, the projects' financial reserves were sufficient to fund the problems experienced. The transferred funds were also not needed for supportability upgrades because they began more slowly than planned due to a longer-than-anticipated approval process. As a result, only about $40 million of the $70 million in funds allocated in fiscal year 1997 to the new supportability upgrade effort will be used; the remainder will carry over to support fiscal year 1998 activities.

The $190 million transferred to the space station program was available because of the large amount of carryover within the shuttle program. Without the transfer, the shuttle program would have had about a $600 million budget carryover at the end of fiscal year 1997. However, shuttle program managers estimated that they would need only $395 million in carryover at that time.

NASA has used carryover to fund upgrades in the past and plans to continue doing so in the future. For example, NASA designated $70 million in fiscal year 1996 carryover funds to initiate the new supportability upgrade effort, and it plans to use another $95 million for this purpose in fiscal year 1998.[2]

Depending on the future upgrades selected, costs could range from hundreds of millions to several billions of dollars. NASA is presently defining an upgrade program to keep the

[2]The $70 million for upgrade activities in fiscal year 1997 included $50 million for supportability upgrades and $20 million for studies of long-term upgrades. For fiscal year 1998, NASA plans to use $75 million for supportability upgrades and another $20 million for continued study of long-term upgrades.

shuttle a viable space transportation system at least through the planned mission life of the space station--about 2013. Other planning assumptions include increasing the shuttle's flight rate from about 8 to 15 times a year, obtaining other customers, and operating the shuttle to 2030. In addition to annual funding requests for safety and performance upgrades and the continued use of excess carryover, NASA plans to use any savings that are generated within the shuttle program to fund its upgrade activities. However, the extent to which these sources will be available in the future is uncertain.

Funding needs for future upgrades will be driven by a number of questions for which answers are not yet apparent. These questions are related to such issues as how long the shuttle will be required and whether viable alternatives can be developed to support space station operations and other human space flight requirements.

FUNDING TRANSFER DID NOT ADVERSELY IMPACT CURRENT AND NEAR-TERM UPGRADE PROJECTS

Transferring funds from the shuttle program had no impact on ongoing upgrades because reserves available in those projects were sufficient to fund the technical problems experienced. Likewise, the funds would not have benefited the supportability upgrade effort because it took longer than planned to initiate those projects.

The four largest ongoing shuttle upgrades represent about 91 percent of the $636 million funding for upgrades in fiscal year 1997. These projects involve the design and development of the (1) alternate high pressure turbopump (AHPTP), (2) super lightweight tank (SLWT), (3) large throat main combustion chamber (LTMCC), and (4) multifunction electronic display system (MEDS). Each of these projects has remained within its initial funding profile because each project has sufficient financial reserves to cover the costs of development problems and schedule delays they have experienced so far. Two of the projects, SLWT and LTMCC, are essentially complete and NASA officials do not anticipate any more development problems. It is still possible that the other two projects could experience further development problems, exhaust their remaining financial reserves, and require additional funding. However, NASA officials believe that reserves would be available elsewhere within the shuttle program if the projects' remaining reserves are insufficient. Table 1 summarizes the cost and schedule status of the four largest upgrades currently underway. Additional information on the status of the four upgrades is in appendix I.

Table 1: Cost and Schedule Changes for Four Largest Shuttle Upgrades

Upgrade projects	Initial funding estimate[1]	Current funding estimate[1]	Funding estimate Change[1]	Initial schedule (month/ year)	Current schedule (month/ year)	Schedule changes (months)
AHPTP	$1,056	$971	$-85	10/97	7/98	+9
SLWT	173	156	-17	12/97	5/98	+5
LTMCC	118	77	-41	10/97	1/98	+3
MEDS	230	206	-24	2/98	1/99	+11

[1]Dollars in millions.

Activities under the new supportability upgrade program have begun more slowly than planned because projects took longer than anticipated for approval. Before work begins, upgrades must be approved through the shuttle Program Requirements Control Board. Proposals to modify the shuttle are thoroughly reviewed through a very formal process before they are approved by the Board. As a result, only about $40 million of the $70 million allocated to the new supportability upgrade effort will be used in fiscal year 1997; the remainder will carry over to support fiscal year 1998 activities.

HIGH CARRYOVER BALANCES ENABLE TRANSFERRING FUNDS AND FUNDING UPGRADES

The $190 million transferred to the space station program was available because of the large amount of carryover within the shuttle program. Without the transfer, the carryover in the shuttle program would have been about $600 million at the end of fiscal year 1997. In contrast, between fiscal years 1993 and 1996, carryover in the shuttle program averaged about $400 million annually. Two factors contributed to much higher than average estimated carryover at the end of fiscal year 1997. First, the $467 million carryover from fiscal year 1996 into 1997 was higher than normal. Second, shuttle program costs in fiscal year 1997 were less than originally estimated. NASA initially estimated program costs for fiscal year 1997 at about $3.1 billion, but its most recent estimate is now almost $100 million less.

In May 1997, shuttle program officials indicated that they would like to have a carryover balance of $395 million at the end of fiscal year 1997–for forward funding on contracts ($200 million), reserves for estimating uncertainties ($100 million), and supportability upgrades ($95 million). NASA's estimated fiscal year 1997 carryover of about $400 million is consistent with the program's average annual $400 million carryover for the four preceding fiscal years. However, it is substantially higher than the $138 million that NASA officials had projected last summer would be carried over from fiscal year 1997 to fiscal year 1998. The $138 million estimate did not include carryover for reserves or supportability upgrades and assumed a much lower carryover from fiscal year 1996.

In July 1996, NASA estimated that its carryover balance in the shuttle program would be about $152 million on September 30, 1996. However, the actual carryover balance on September 30, 1996, was over $467 million. The large increase was due to NASA not using all of its program reserves ($122 million), underspending or underrunning costs in some program elements ($185 million), and reserving funds to cover costs associated with the transition to a single space flight operations contract for the shuttle ($19 million).

NASA has used carryover to fund shuttle upgrades in the past and plans to continue doing so in the future. NASA designated $70 million in fiscal year 1996 carryover funds to initiate a new effort involving supportability upgrades in fiscal year 1997, and it plans to use another $95 million for this purpose in fiscal year 1998. Beginning in fiscal year 1999, supportability upgrades will be funded as a part of the safety and performance upgrade portion of the shuttle program's budget. Some of the supportability upgrades approved so far include upgrading the checkout and launch control system. Funds are also being used to study potential long-term upgrades such as liquid flyback boosters.

ISSUES WARRANT CONSIDERATION BEFORE FUNDING FUTURE UPGRADE REQUIREMENTS

NASA is presently defining an upgrade program to keep the shuttle a viable space transportation system at least through the planned mission life of the space station, about 2013. Additional planning assumptions include operations until 2030, a flight rate of 15 per year, and use of the shuttle by both the Department of Defense and commercial customers.

The potential funding requirements for future shuttle upgrade activities could be substantial. NASA has identified about 70 potential modifications to the shuttle with an estimated total cost of $5 billion to $7 billion. Most of this estimated cost relates to development of the liquid flyback booster. Over 30 upgrades with an estimated total cost of about $322 million have already been approved for implementation or study.

NASA plans to use a variety of sources to fund these activities, including the safety and performance upgrade budget line, the continued use of excess carryover balances, and savings from the new space flight operations contract. Shuttle program officials are depending on the last two sources to build an account to help fund future upgrades. However, the extent to which all of these sources together will provide sufficient funding for upgrades is uncertain. For example, NASA's overall future budgets are currently planned at slightly reduced levels, but larger reductions may occur. Also, the continuing availability of sufficient excess carryover balances is questionable, and the level of any future savings under the space flight operations contract is unknown. Ongoing NASA studies, including the one directed by OMB, are expected to provide additional information about the shuttle's role in the Nation's space transportation strategy. The results of these studies will help initiate the necessary debate for shaping the U.S. government's future space transportation policy.

Funding needs for future upgrades will be driven by a number of questions for which answers are not yet apparent. These questions are related to such issues as (1) how long the shuttle will be required, (2) what upgrades will be needed to maintain the shuttle's safe operation, (3) whether there will be any savings from the space flight operations contract and excess carryover, (4) who should pay for required upgrades if the shuttle is privatized, and (5) whether there are viable alternatives to the shuttle to support space station operations and other human space flight requirements.

\- - - -

Mr. Chairman, this concludes our statement. We would be happy to answer any questions that you or members of the Subcommittee may have.

STATUS OF THE FOUR LARGEST ONGOING SHUTTLE UPGRADES

The National Aeronautics and Space Administration's (NASA) four largest ongoing upgrade projects involve the design and development of the (1) alternate high pressure turbopump (AHPTP), (2) super lightweight tank (SLWT), (3) large throat main combustion chamber (LTMCC), and (4) multifunction electronic display system (MEDS).

Alternate High Pressure Turbopump

The AHPTP project is designed to upgrade the shuttle main engine's high pressure oxygen and hydrogen turbopumps. The turbopumps inject liquid oxygen and hydrogen fuel into the engine's combustion chamber where they mix and burn to generate power for shuttle launch. The oxygen turbopump has completed development and is operational on the shuttle. The hydrogen pump development was suspended in fiscal year 1992 and restarted in fiscal year 1994. The hydrogen pump is currently scheduled for its first flight in July 1998.

NASA officials told us that the technical difficulties associated with the hydrogen pump are due to design problems related to cracks in the housing and turbine blades. At this point, the solution is to redesign and retest to verify that the problems are resolved. According to NASA officials, currently available or planned funding reserves are adequate to cover any additional costs. As a result, NASA officials said that additional funding would not benefit the project.

Super Lightweight Tank

SLWT will provide the shuttle with additional lift capability to support space station assembly and operation. The tank is built from a new aluminum alloy that reduces the tank's weight and provides the shuttle with an additional 7,500 pounds of lift capability. SLWT is currently scheduled for its first flight in May 1998.

SLWT technical problems have been primarily associated with welding the new aluminum alloy. When welds intersect, cracks can occur which cause time consuming repairs. According to NASA officials, the problems are related to the manufacturing process, and solutions have been identified and implemented. SLWT has completed development and has been certified for flight. The first flight tank is currently being manufactured.

Project officials believe there are sufficient reserves to fund any unforeseen problems. The SLWT project completed development with about $20 million in reserves with another $12 million in reserves at headquarters.

Large Throat Main Combustion Chamber

Liquid oxygen and hydrogen fuel mix in the space shuttle main engine combustion chamber to generate the engine's power. LTMCC, which is an advanced design of the current combustion chamber, will reduce the operating temperature and pressure within the engine and thereby increase the engine's safety margins. LTMCC is on schedule and within estimated cost and is currently scheduled for its first flight in January 1998.

Multifunction Electronic Display System

The space shuttle orbiter currently provides flight control information to crew members through cathode ray tube displays, instruments, and meters. The current system consists of hardware based on 1970s technology, which creates ongoing logistics support problems because of wear, aging, and component obsolescence. MEDS is being designed to simplify orbiter operations, increase the life of the orbiter's displays, and increase the reliability of the system. MEDS is currently scheduled to become operational in January 1999.

MEDS has a development problem with the glass required for the displays. NASA attempted to develop a U.S. supplier for the glass. However, the supplier chosen was subsequently unable to meet program production requirements and could not meet the stringent quality standards for the severe operational environment of the orbiter. As a result, the U.S. supplier's contract has been terminated, and the glass is being purchased from a foreign source.

NASA estimates that obtaining a new glass supplier will add about $20 million to the cost of the project, but reserves have thus far been adequate. However, contract negotiations for the new glass supplier are not complete, and the contractor's proposal is $6.8 million more than available in program reserves. NASA officials told us that they hope to resolve this issue through contract negotiations. The officials said that, if negotiations are not successful and the increased cost materializes, they will use funds within the shuttle program to meet the additional funding requirement.

(707289)

Chairman ROHRABACHER. Thank you very much.

Last, but not least, we have Jim Adamson, the Chief Operating Officer for United Space Alliance, NASA's space flight operations contractor. I met with Mr. Adamson, and we've met and spoken before, and we're very pleased to have him with us today to give us his first-hand report.

Mr. Adamson.

TESTIMONY OF JAMES ADAMSON, CHIEF OPERATING OFFICER, UNITED SPACE ALLIANCE

Mr. ADAMSON. Thank you, Mr. Chairman and members of the Subcommittee. I would like to thank you for the opportunity to address this panel especially on the subject of Shuttle safety. I'd also like to thank Mr. Weldon for his kind remarks about our workforce at the Kennedy Space Center. We're very proud of our employees and we do believe that we have the best launch team in the world.

The topic is timely today in that October 1 is the first anniversary of the SFOC contract, and you do have my written statement, and I would like to request that it be made part of the record.

Chairman ROHRABACHER. Without objection, and that is true for the rest of the testimony as well.

Mr. ADAMSON. Let me say at the outset that the United Space Alliance has felt no pressure whatsoever to reduce cost or accelerate savings at the expense of safety. The subject funds transfer has had absolutely no impact on the SFOC contract.

Last Thursday we launched a multi-national crew aboard Atlantis on STS86, the 8th Space Shuttle mission to be launched by the United Space Alliance in its first year of operation. By now you may know that we did not launch exactly on time. We were 58 milliseconds late against a launch time that was set months before.

I think that our record speaks for itself. During the past 12 months, 7 of 8 flights were launched on schedule according to the performance standards of our contract. Only one flight was delayed for one day to better insulate a waterline in the payload bay. The only other impacts were minor impacts due to weather and range scheduling constraints. In fact one flight was actually accelerated 2 days in order to accommodate the range scheduling.

In flight anomaly average is down to about 5 per flight from an average of about 14 over the previous 2 years, and I think that Mr. Trafton would tell you that they are down even more than that from several years before 1991–1992 timeframe. I think that we're down about 50 percent.

With the exception of STS80 during which the external airlock hatch failed to unlatch, all major mission objectives have been accomplished or exceeded. A reflight was required for STS83 which was cut short after 4 days out of concern for the performance of a fuel cell. In both cases, United Space Alliance participated actively in the management decisions and strongly recommended the conservative approach in the interests of safety.

The STS83 crew was brought home safely, the fuel cell was replaced, and after a record 84 day turn around, Columbia was relaunched with the same crew and the same set of experiments aboard STS94. It was the first such reflight in Shuttle history. A test case for quick turn arounds that might be required during space Station construction.

During the course of the mission, all of the objectives were met or exceeded, and only three inflight anomalies were identified.

In addition to flying Space Shuttles safely and reliably, I am pleased to report that the ongoing transition of Space Shuttle operations tasks, processes and management responsibility from NASA to the United Space Alliance is on track and proceeding well. Three hundred and seventy four tasks have been scheduled for transition to date, and 100 percent of those have been transitioned.

Over the last year, we have created a new company with new business and management systems consolidating successfully the work content and administration of 12 different contracts. We have exceeded all major goals associated with OSHA injury rate, government property damage type A and B mishaps. We achieved an international standard certification, ISO9000 for NASA Shuttle Logistics Depo Flight Operations and Ground Operations ahead of schedule.

Throughout the transition, United Space Alliance has exhibited a strong commitment to safety and mission assurance. Each and every step has been closely examined to ensure that no additional safety concerns are introduced. USA has adopted a best practices model for safety and risk management, and before any changes to that model are made, we will coordinate and review extensively with NASA. We are working continuously to reinforce our "Safety First" culture through such initiatives as ISO9000 certifications, the Safety Hotline and an upcoming Safety Awareness Day. This kind of safety and reliability is indicative of a healthy program. I am proud to say that after 1 full year under the SFOC contract the safety of the Shuttle program has never been better.

Thank you, Mr. Chairman, that concludes my remarks.

[The prepared statement of Mr. Adamson follows:]

USA
United Space Alliance

Hold for Release Until
Presented by Witness

Jim Adamson
Chief Operating Officer

1150 Gemini
Houston, Texas 77058-2708
Telephone: 281.212.6120
Fax: 281.212.6179
jim.c.adamson@usahq.unitedspacealliance.com

Statement of
James C. Adamson
Chief Operating Officer
United Space Alliance

before the

U.S. House of Representatives
Committee on Science
Subcommittee on Space and Aeronautics

Hearing on Space Shuttle Program Safety
October 1, 1997

Statement of
James C. Adamson
Chief Operating Officer, United Space Alliance
before the
U.S. House of Representatives
Committee on Science
Subcommittee on Space and Aeronautics
Hearing on Space Shuttle Program Safety
October 1, 1997

Mr. Chairman and Members of the Committee.

Introduction

I would like to thank you the opportunity to address this panel on the subject of the safety of the Space Shuttle Program. I am very pleased to report that the ongoing transition of Space Shuttle operations tasks, processes and management responsibility from NASA to United Space Alliance (USA) is on track and proceeding well after one full year under contract, and that the safety of the Shuttle Program has never been better.

Over the last year, we have created a new company with new business and management systems, consolidating successfully the work content and administration of 12 different contracts into a single contract. In the process, we have restructured two major subcontracts, achieving a savings of well over $100 million over the length of the contract.

We successfully have completed the first Orbiter Maintenance Down Period under the contract, returning Endeavour to the fleet under cost and ahead of schedule. We have exceeded all major goals associated with Lost Time, OSHA Injury Rate, Government Property Damage and Type A&B mishaps. We achieved the International Standards Certification (ISO 9001) for the NASA Shuttle Logistics Depot, Flight Operations and Ground Operations ahead of schedule.

We are working continuously to reinforce the "safety-first" culture of the program through such initiatives as ISO 9001 certifications, a safety "hotline" and employee Safety Awareness Days.

Throughout the transition, United Space Alliance has exhibited a strong commitment to Safety and Mission Assurance. Each and every step in the transition is closely examined to ensure that no additional safety concerns are introduced by the transition. USA has adopted as a baseline the post-Challenger model of safety procedures, checks and balances. Before any changes to that baseline are made, we will coordinate and review extensively with NASA.

The First Year: Safety and Performance of the Shuttle Team

Since the start of the Space Flight Operations Contract (SFOC) just one year ago, we have safely and successfully launched eight and landed seven Space Shuttle missions on schedule. We expect to add another perfect landing when Atlantis returns - next week - after another successful docking and re-supply mission to the Russian space station Mir.

With the exception of STS-80, during which the external airlock hatch failed to unlatch, all major mission objectives have been accomplished or exceeded. A re-flight was required for Shuttle mission STS-83 which was cut short after just four days, out of concern for the state of a flight-critical fuel cell. In both cases, USA actively participated in the mission management decisions and strongly recommended the conservative approach in the interests of safety.

The STS-83 crew was brought home safely, the suspect fuel cell was replaced and, after a record 84-day turnaround, Columbia was re-launched with the same crew and the same set of experiments aboard STS-94. It was the first such re-flight in shuttle history, a test case for quick turnarounds that might be required during space station construction. During the course of the mission all objectives were met or exceeded and only three in-flight anomalies were identified.

Our on-orbit performance record is something of which we are particularly proud. The average number of in-flight anomalies recorded during each flight under the SFOC represents a thirty-five percent reduction in anomalies recorded the previous year…that, coupled with the schedule performance of the Shuttle team is solid evidence that program has never been more safe.

We are also proud of the fact that, while Shuttle safety and reliability is at an all-time peak, we have also remained on track with our goal to save the U.S. taxpayer in excess of $400 million during the life of the contract.

The Transfer of Funds to the International Space Station

During the last year, a decision was made by NASA to transfer $190 million from the NASA Shuttle budget to the International Space Station (ISS) budget. It has been our understanding that the funding was volunteered by the Shuttle Program from reserves built up as a result of management and planning efficiencies. United Space Alliance has never been asked to spend or accept less funding than has been budgeted for the Space

Flight Operations Contract as a consequence of that decision. The contract baseline remains the same.

In support of that baseline, USA remains strongly committed to processing and operating the Space Shuttles as efficiently as possible without compromising safety. The terms of our contract with NASA drive safety to the top of the agenda through both direct safety assessments and fee determination.

We have developed a quantitative risk assessment plan with the primary goal of identifying and recommending specific management and organizational strategies which will enhance safety and optimize Space Shuttle performance. More than an operational safety tool, this risk-based decision-making approach is intended to enhance overall business performance, provide an objective evaluation of all possible risks, and quantify benefits and costs before implementing new strategies.

The formula for determining our fees includes unique performance fees for achieving mission objectives as well as significant penalties for failing to meet safety and mission requirements. This effectively ties financial success directly to the safety and success of the program. Every element of our fee structure must pass the safety test before it can be earned.

The contract also includes fee incentives to reduce the cost of the program. But it is important to understand that profit is not the sole motive for reducing cost. USA views these incentive earnings as opportunities to redirect funds to key Shuttle upgrades which will keep the program safe and viable for the long term.

Replacement Parts and Vendor Support

Performance under the SFOC, in terms of replacement parts on-time delivery has been outstanding, meeting or exceeding maximum performance standards in every specified category. While maintaining the highest performance standards, we are also working to mitigate the risk that exists in the program relative to obsolescence and supplier loss.

As our Shuttle fleet matures, the risk of hardware obsolescence and loss of original suppliers will naturally increase. USA is actively mitigating this risk through improved supplier communication, technology upgrade strategies, and the development of in-house manufacturing capabilities.

A supplier council has been established to improve understanding of the unique operating environment of the Shuttle through improved communication, education, and motivation. In addition, a subcommittee has been formed to develop a process to assist each supplier, which has agreed to achieve ISO 9001 certification.

We are continually evolving the manufacturing capability of NASA's Shuttle Logistics Depot to absorb the loss of unique parts suppliers when new suppliers cannot be found.

Driving directly to the heart of the obsolescence issue is technology. In that regard, we are working to identify Shuttle upgrade candidates which employ current technologies for which there is an ample supplier base.

Shuttle Upgrades

USA is very pleased to be participating fully with NASA in the enhancement and upgrading of the Space Shuttle. We applaud the NASA management position to solicit and consider USA input in the planning and prioritization of potential upgrades.

USA expects to be operating the Shuttle well into the next century with the goals of flying safely, meeting the manifest, and reducing cost. We believe that the upgrades program must be focused on these three fundamental goals. The USA upgrades team is currently evaluating and ranking all candidate upgrades against those goals in support of specific recommendations to the customer.

In addition to these criteria, USA is also assessing the upgrades to identify specific candidates for investment using program savings and private funding. Such upgrades would include those which significantly enhance performance and reduce operating cost.

In terms of ongoing upgrades, USA is continually assessing minor vehicle and ground equipment modifications as potential investments in overall operating efficiencies. We have already invested our own funding in improvements such as the orbiter body flap cove refurbishment and the modification of the Solid Rocket Booster (SRB) retrieval vessels to tow the External Tank (ET) barge, which is used to deliver the ET to the launch site for processing. These innovations not only represent cost savings to the program over time but the enhancement of valuable government-owned property.

Effects of Delay in ISS Assembly on Shuttle Manifest

Although the delay in the start of Space Station Assembly will result in one less flight on the manifest in 1998, the schedule slippage will be made up by Assembly Flight 7A in 1999. We do not anticipate any unmanageable effects or stresses in ground or flight operations as a result. In fact, it is our belief that there is considerable excess capacity in the projected manifest that will serve the ISS Program well in terms of flexibility and supportability.

In support of a recent joint study by USA and NASA on the costs of performing major modifications to the Shuttle Orbiters, an intensive analysis of launch site vehicle processing logistics was conducted. That analysis determined that, by implementing achievable efficiencies, Orbiter processing turnaround time could be significantly reduced, further enhancing manifest flexibility as well as creating the opportunity to safely increase the flight rate.

By increasing the flight rate and spreading the fixed operating cost over a larger number of flights per year, the incremental cost per flight could be significantly reduced. The addition of commercial revenue-producing flights to the manifest would even further reduce the overall cost of the program to American taxpayers, and contribute to a safer overall program. It is our sincere belief that a stable operation at a healthy flight rate will boost employee morale, help to maintain currency and proficiency and, as a direct result, contribute to maximum program safety.

Commercial Payloads & Shuttle Privatization

This week, we are delivering a draft Shuttle Privatization Development Plan to NASA, which proposes an approach to flying commercial payloads within a year and privatizing the Shuttle by the year 2002. We applaud NASA for the vision to request such a plan as a deliverable product under the terms of our contract.

The plan recognizes that there are many issues to address on the road to privatization - the legislative and policy barriers that presently exist, the development and stimulation of a viable and appropriate commercial market, the achievement of program efficiencies and investments that will be necessary to support an increased flight rate. This plan offers a process for collaboration between USA and NASA to address and resolve these issues.

We believe the time is right to address these issues and define the role of the Shuttle in the commercial arena. The needs of the global space marketplace have only begun to catch up to the capabilities of the Space Shuttle - fortunately, at the same time that the Shuttle program has reached a level of significant stability and maturity.

We are already joining with commercial aerospace companies, like Spacehab, to identify and develop a viable commercial market for the unique abilities of the Space Shuttle. Although the commercial launch services market is not presently accessible to the Space Shuttle, significant demand does exist to support a role for the Shuttle in that market - most likely a complementary rather than a competitive role.

Ultimately, the unique capabilities of the Shuttle, which distinguish it from other space launch systems will define it's place in the commercial marketplace. Payloads and research customers which stand to substantially benefit from human operations and intervention in space will be a natural market. The areas of life sciences, material sciences, space observation, education and entertainment all offer future opportunities in space. Interest exists in on-orbit assembly, retrieval and repair, the development of commercial manufacturing facilities, and support activities associated with the assembly and operations of the ISS. In tandem with a commercially developed space tug, the Space Shuttle would enable existing equipment to be salvaged, and delicate equipment to be delivered and assembled in space.

Reliability and cost will ultimately determine the Shuttle's success in the commercial marketplace. The Space Shuttle fleet is the safest, most versatile, reliable, and successful space launch platform in the world. The ongoing consolidation and integration of contracts under the SFOC, improvements in Orbiter processing and modifications, key Shuttle upgrades and a continual commitment to safety will continue to improve the cost effectiveness of the program.

The cost per pound to orbit will no longer be the sole cost/benefit determiner in the marketplace. The value added by the Shuttle's operational capabilities and human intervention…the investment value of satellites retrieved in orbit for repair or return...the economies of scale involved in multi-disciplinary, multi-customer missions will all factor into the overall value and success of the Shuttle in the commercial marketplace.

The Shuttle provides payload delivery range and operational capability, including human intervention, that other systems won't match for many years. In the meantime, the doors to commercial enterprise in space can be held open by a Shuttle program that is fully utilized and delivering maximum value.

In short, it is our vision to bring the practical benefits of a stable, continually maturing program down to Earth, as our primary customer - NASA - continues it's remarkable exploration of the universe in which we live.

###

James C. Adamson
Chief Operating Officer, United Space Alliance
Biographical Sketch

Jim Adamson has served as USA's chief operating officer since the company's inception in November 1995.

Prior to joining USA, he held key executive management positions with Lockheed, including Group Vice President and General Manager of Lockheed Martin Engineering and Science Services and President and CEO of Lockheed Engineering and Sciences Company. Previously, Mr. Adamson was a two-time Shuttle flyer as a NASA astronaut.

Mr. Adamson was commissioned a Second Lieutenant in the Army upon his graduation from West Point in 1969. He served as a military test pilot logging more than 3,000 hours of flight time.

Selected by NASA as an astronaut candidate in 1984, Mr. Adamson was assigned to the flight crew of STS-28, a Department of Defense mission flown in August 1989. He was also a member of the STS-43 crew on a mission to deploy the Tracking and Data Relay Satellite (TDRS-E) in 1991. Having logged more than 334 hours in space, Mr. Adamson retired from government service in 1992.

Chairman ROHRABACHER. Thank you very much, Mr. Adamson, and we are pleased to hear, of course, that Shuttle safety has not been compromised.

I for one am just a little bit astounded about the amount of money, that, of course, Mr. Li described in detail for us, that seems to be floating around as if weightless in space and not really directed by Congress towards specific purposes in America's space program. We're talking about, as Mr. Li described, hundreds of millions of dollars. Am I not correct Mr. Li? Hundreds of millions of dollars that are just there and at sort of discretion—not just discretion on how to use it for a specific program, but it now seems that we are being told that it is discretion within various programs.

Mr. LI. Last year we testified before this Subcommittee about the entire NASA budget in the context that I referred to today, about the past estimates of how much carryover would be created—and NASA will I'm sure want to discuss this issue. But the components of carryover can be in many facets. Yes, there are some reserves for funds that are unobligated. However, there are also some situations, Mr. Chairman, when you have costs, that the bills have not yet come in. Funds are obligated, but they are just uncosted at that point in time. So, when we discuss the issue of the $600 million carryover, I don't want to give you the impression that these are funds that are totally available for any program.

Chairman ROHRABACHER. I see. Because, Mr. Trafton, during your remarks you said that the Shuttle will do its part when you were talking about the various needs for the space Station, etc., I guess. Is that what you are referring to?

Mr. TRAFTON. Yes, sir.

Let me say a couple of words about uncosted as well. No program manager in his right mind will come to the end of the fiscal year without some uncosted. If my Shuttle program manager came to me today and said, "You'd be proud of me. I had zero uncosted on September 30," I would fire him. You just can't run a program that way for reason——

Chairman ROHRABACHER. Okay. I can appreciate that, and I understand what that means. My father was a Marine Lieutenant Colonel, and he commanded a fighter squadron, and at the end of the fiscal year he would have his planes—now my dad is very patriotic and as frugal a person—he comes from North Dakota—as any person that I know. But at the end of the fiscal year he would have his planes flying around in circles so that he would use up all the money so that it would be allocated resources.

Now we don't want NASA to be put in that situation where it feels that to get the same allocation it has to use up all of the funds that it gets from us every year. That's wrong because it encourages people to just be frivolous with their spending. I understand that. That's not what we're saying here.

However, we do get to a point—and I don't want to get too personal here—but when you say the Shuttle will do its part, the Shuttle isn't doing its part. It's the taxpayers that are doing their part, and we're trying to make sure that we understand here that these are taxpayers dollars that we're talking about and it is not the purview and the right of any government official to just direct those dollars where they see fit.

But the point you just made is certainly valid. We don't want to put you in the spot where at the end of the fiscal year you are going to be using it up just to have used it up. That would not be good management on our part.

However, I do want to warn about the attitude that this is Shuttle money. It is not Shuttle money. It is taxpayers' money.

With that, I have a couple of other questions of course, but I think that we will let Mr. Cramer and some of our other members ask theirs first.

Mr. Cramer.

Mr. CRAMER. I would yield to my colleague from Texas. He may have to leave a little early. I want to give him an opportunity to talk.

Mr. LAMPSON. Thank you, Mr. Cramer and Mr. Chairman.

I am just curious, Mr. Li, what is the percentage of the holdover to the total budget that is not costed? On average.

Mr. LI. Currently—let me give you some background about the carryover amount that NASA is currently projecting. And today is the beginning of the fiscal year, but there are some bills that are just not in yet.

They are projecting about $400 million. Their budget authority is slightly under $3 billion. So—the percentage is not in the double digits.

However, the issue that I have been making in both my statement and that we made last year in the statement is that each single program has different needs and they are unique. An operational system like the Space Shuttle probably has a different quantitative level of carryover that is needed versus an R&D program. An R&D program would have to have a higher percentage. So the percentage would be the $400 million versus the $2.9 billion.

Mr. LAMPSON. Well I have been managing a fairly good-size budget for almost 20 years and I know that you cannot do it without having some carryover, without having some plan for miscellaneous or unexpected expenses or you get yourself into some serious trouble. Just curious about that particular point.

Mr. Trafton, does the delay in the assembly schedule for the International Space Station pose any kind of a threat to the Shuttle Program's safety due to changing loss requirements?

Mr. TRAFTON. Not at this time, no, sir.

Mr. LAMPSON. And has the reallocation of funds from the Shuttle program to the ISS program have any effect on ongoing upgrade programs or to the schedule for future upgrades?

Mr. TRAFTON. It has not.

Mr. LAMPSON. What is the expected life of the Space Shuttle? How much longer might we be able to expect it to continue to be used as effectively as what it has been?

Mr. TRAFTON. With the currently planned upgrades, we hope that the Shuttle—it is our plan that the Shuttle will be around to the completion of operations on the International Space Station. So we're talking 2012, 2013 time frame.

Mr. LAMPSON. Good.

Keep working on your time records, Mr. Adamson. It sounds like you are very proud of what you have done, and 58 milliseconds—maybe you can improve on it. Bring it on down to 30.

Mr. ADAMSON. It's not too bad. Thanks.

Mr. LAMPSON. That's all, Mr. Chairman. Thank you.

Chairman ROHRABACHER. Mr. Weldon.

Mr. WELDON of Florida. Thank you, Chairman.

Let me follow on that theme that my colleague from Texas was talking about. I think that we were all very impressed with the rapid turn around on STS83, I think it was, which was the micro-gravity experiment.

My question to you, Mr. Adamson, is whether some unique problems you encountered in that you were able to overcome, and is it possible to turn around the Shuttle even more quickly than what you did in that particular situation. The direction that I am heading in all of this since I know that your company, USA, is looking at the prospect of privatizing and the possibility of perhaps in the future of lifting the White House Executive Order on restricting the Shuttle on having commercial payloads—I guess my question there is could you just give us a brief overview of what are your goals and what is your intent to accomplish there. And I know that maybe we will get into this in more detail in the future, but perhaps maybe you can briefly talk about the possibility of commercializing the Shuttle again and putting commercial on it, or DOD payload.

Mr. ADAMSON. Well, first of all, sir, with respect to STS83, turning that flight around. It was a unique opportunity, but let me add that a turn around schedule like that could not be sustained in an ongoing operation. In that particular case we had a trained crew, a ready payload set up in the bay. There are many variables in turning the flight around and—so all of the piece parts and the materials and training levels were all in place to make that happen. So, not wanting to take anything away from the processing team, it was set up to be done quickly because they had just returned. Basically all we had to do was pull the fuel cell out and replace it and that also initiated an upgrade to the fuel cells which allowed us—which will allow us to improve the instrumentation so that hopefully we can prevent that particular problem from recurring.

With respect to turn around time and privatization of the Shuttle and any future commercial—potential commercial market for Shuttle, the turn around schedules and the processing flow times really need to be made consistent with the flight rate. It has always been our goal to try to level the resources in the Shuttle program as much as possible, so we like to prevent peaks and valleys of work intensity from occurring. To the greatest extent that we can distribute the manifest and keep that a steady flow, the better we're able to level our resources and keep everybody busy. You don't want to staff up for the peaks and then have to suffer long periods of valleys. So we're constantly struggling with that and trying to level that and balance that against the manifest requirements that NASA has for the flights.

As we go down the road, it is our desire to further to Mr. Trafton's carryover problem by making the Shuttle program cost less and less. There are two reasons for that.

One we would like to make the Shuttle as competitive as possible so that we do have an opportunity to add commercial flights and therefore help offset some of the taxpayer dollars with an infusion of commercial revenues. We hope—that's our dream that some day we'll be able to do that with the Shuttle and make that excess capacity available on the commercial market place and international market.

To that extent we will continue to work to make the operation more efficient and reduce the flow times in the Shuttle program. It is the chicken and the egg argument. The more we can become efficient and reduce our costs, the more competitive we will be, the more customers we can hope to attract, and the more customers we hope to attract the more flights we can add. And certainly flight rates in the range of 9 to 10 to 12 are well within reach of the program as we see it today. When we start having demand for that kind of a flight rate we will then have further demand to reduce our flow times, and within the confines of keeping safety first and not letting anything slip through the cracks with regard to reliability, we will continue to work to that end.

Chairman ROHRABACHER. Thank you very much.

Mr. Cramer?

Mr. CRAMER. Thank you, Mr. Chairman.

My initial questions are for Mr. Johnstone. How often does your advisory panel review the Shuttle program? What do you look for in your reviews, and what are your top safety concerns?

Mr. JOHNSTONE. Within the limitations of the time that we are alloted by law, I would say that we are continuously looking at the Shuttle program. We make four or five trips a year to Cape Kennedy to see how it's going down there. We talk to the people at Johnson and Marshall and at headquarters probably on an average of twice a year as a full panel.

I think that is probably about all we need to do. It keeps us pretty well updated on whatever is going on in the whole Shuttle program.

Mr. CRAMER. What's your top safety concern?

[Long pause.]

Mr. JOHNSTONE. It is sort of complex because it gets into the change that has gone on at NASA within the last few years. Some of it——

Mr. CRAMER. Organizationally?

Mr. JOHNSTONE. And we hear that some of it is going along very nicely. There are small pockets of places where maybe the staff has been reduced faster than NASA expected, and they are a little on the backside of the power curve in terms of manpower. There are not huge areas of concern, but it is just the total change and turmoil that comes with all aspects of it. We don't normally look at money. We try to stay out of money because in our opinion that has nothing to do with safety.

Mr. CRAMER. So would you say——

Mr. JOHNSTONE. Unless you cut money too far, and then you've got a problem.

Mr. CRAMER. Are you something then—more then something, you are a watchdog over how they reinterpret themselves as they change organizationally?

Mr. JOHNSTONE. Absolutely. In fact, we've been referred to that because of our Congressional mandate that we have a license to meddle, and we don't mind doing it.

Mr. CRAMER. We want you to do that often and very well.

You reiterated your concerns over Shuttle obsolescence issues. Would you elaborate on that a little bit.

Mr. JOHNSTONE. Well there a lot of areas. The Shuttle is a peculiar operational system if you will. There are four of them flying. The production line has ended. You've got subcontractors and sub-subcontractors. They are not at all interested in this business anymore. They are out of it really. And to keep up the spares supply, and then you drop on top of that a lot of elements of the Shuttle that no longer meet the environmental regulations that exist in the country. And how do you get that out of the Shuttle or into—out of the system when it was perfectly legal 25 years ago when it was built but it is not legal anymore and you don't have contractors that are overly happy about having to go do that. It is not like you were building a triple seven where you are looking at a 400 airplane production line and a long run of profit off of that production line.

Mr. CRAMER. Mr. Trafton, I want to give you the opportunity to comment about—we have just finished our one year anniversary of the United Space Alliance. How has that gone? What would you say to us? Does anything else need to be done—we were very skeptical, or at least cynical to a certain extent about what you would accomplish and how quickly you would accomplish it. Would you just give us kind of your reaction to this one year anniversary.

Mr. TRAFTON. I think that it has gone very well sir. We had a couple of problems, but we overcame them. There was STS80 and 83 both had problems. The USA company reacted very well in both cases. We're moving forward with the transition, the phase two transition. We have 12 contracts under them now. We're talking to them about another 16 contracts here in the near future. We hope—also today we are getting a deliverable on the contract which is their view of a privatization plan for the Space Shuttle Program. We'll take a look at that and digest it within NASA and we'll talk to the Administration, and of course we will also discuss it with this Subcommittee and your staffs.

So overall I think that things are going very well.

Mr. CRAMER. So you are accomplishing everything that you hoped you would accomplish?

Mr. TRAFTON. That is correct.

I will say, though, that in terms of dollar savings, I think that we're kind of flattened out right now. I don't think that we're going to see any large dollar savings through the SFOC contract through the next couple of years. I think that's down the road a ways.

Mr. CRAMER. Thank you.

Thank you, Mr. Chairman.

Chairman ROHRABACHER. Thank you, Mr. Cramer.

Just a note from Mr. Johnstone's testimony. We're talking about 1960's and 1970's technology in the Shuttle. This is—what you talked about and what your major concerns seem to be is inherent in trying to use the technology for the number of decades that we have been using the Shuttle, and especially when you are talking about subcontractors and contractors that no longer want to be in the business because they have moved on to other things. Basic piece of technology is old—the Shuttle. This is not new technology.

And the upgrades that we are talking about hopefully will make it safe until we can move on to a newer technology and bring a newer team in and newer people who were involved with subcontracting the parts which would take care of that.

Let me just note that there are people in Washington who want to see the Shuttle program be the basis of America's space transportation system until 2030. We've got people who have great plans for the Shuttle. And every problem that you've talked about and that you spoke about—and it's not serious now, but it is something that you have to be very careful about and watch—will be exacerbated if we try to extend the life of a piece of technology that was not meant to last this long. Not to say that it is going to wear out, and not to say that the upgrades are not going to make things better. But the fundamental concern that you have will get worse if we begin thinking that we're going to rely on the Shuttle rather than develop new systems. That's why this Chairman has made his number one—I don't want to talk in the third person—that is why I have made my number one goal for the last 5 and 6 years to try to develop a new reusable rocket system, or at least a new system of technologies on which we can depend upon as a space transportation system.

Mr. Cramer is shaking his head—it is a bi-partisan concern. So just to put it into perspective—your testimony in perspective—what's come out of this is that your main concern for safety is this general uneasiness that you feel about that out there. I think that it was well said and that we should heed that warning.

Yes, sir?

Mr. JOHNSTONE. I think that's a good interpretation, Mr. Chairman.

Let me switch over to my career experience, and I might carry some of you back a little too far. But how long was the DC3 supposed to be flyable and safe? And yet we kept improving, we kept improving.

Chairman ROHRABACHER. Right.

Mr. JOHNSTONE. If we had said that the supersonic transport was going to be the airplane for the future, we would have quit building subsonic airplanes. And where would we be today if that had happened?

The point with the Shuttle, I believe, is that until you've got in your hand a replacement vehicle that can do the job better than the Shuttle, then you better keep the Shuttle in top-notch shape.

Chairman ROHRABACHER. Right.

Mr. JOHNSTONE. And you don't just say, "Well, we're going to replace it in 2 years," because you are not going to replace it in 2 years. So you better give those guys who are flying it the best technology that is available and keep that Shuttle going as long as it is needed.

Chairman ROHRABACHER. I think that is an excellent point. I think that I would say that if we spent all of our money upgrading the DC3 and did not move forward with the 707 with the money necessary—because it required some money to develop the 707. It was done in the Air Force, I understand. That would have been a bad decision as well. Up until recently we have not been putting our resources into developing the next step. And we have to make sure that we are not just improving our current system to keep it flying at the expense of developing the next step.

Mr. Nethercutt?

Mr. NETHERCUTT. I thank the Chairman, and welcome, gentleman. I thank you for your testimony.

Mr. Trafton, I want to ask a few questions of you, sir, about what the Chairman referred to as money floating around and reprogramming, in particular from the Shuttle safety and performance upgrade program to the International Space Station.

In April of this year, I looked at your testimony in which you expressed an interest to this Committee in reprogramming about $200 million for the International Space Station. And my review of your testimony is that, you stated, "We do not intend any 1998 or beyond dollars in the Shuttle program, and this will not impact Shuttle safety." Do you still stand by that, sir? Do you intend to do anymore, or do you expect the Agency to do any more reprogramming to cover space Station shortfalls?

Mr. TRAFTON. Yes, sir, we're going to have to do some things in 1998. We're talking about $50 million right now, but I think that as we get our work plan in place for the Shuttle, and we come out with an operating plan for the Shuttle and the Station, there may be some additional funds required. It is just, as I've stated earlier, so absolutely important to us that we keep the space Station on schedule. If we don't do that then the Shuttle doesn't have the payloads to fly and the Shuttle quickly drops below what we think is a safe flight rate which is about five flights per year.

And with regards to transfers, sir. We don't just sit around and write checks to each other here at NASA headquarters. We have gone through the right processes here in transferring funds between programs.

Mr. NETHERCUTT. I appreciate that. My concern is with safety and whether we're transferring money, reprogramming money to the space Station at the expense of the safety of the Shuttle.

As Mr. Johnstone has said, we have to keep the Shuttle in top-notch shape, just moments ago. And I don't think that we can continue to defer upgrades on the Shuttle, personally, because I think that safety is what we have in mind. I know that this Committee has fully funded the safety upgrade requests of the Administration in the past for years especially.

I also want to know, sir—I assume that the POP97 Shuttle program recommendations, you are familiar with. It is a serious planning document, I assume. I don't know if you had any involvement in preparation of it, but I would refer you to page 10 of the POP97 recommendations in which the Space Shuttle Program is stated, "The Shuttle upgrade requirements were and continue to be well in excess of $100 million a year. Our plan was and continues to be to fund additional upgrades from operational underruns in the program. Fiscal Year 1996 underrun was a start toward that end." And then this comment leapt out at me, "A hundred and ninety million Fiscal Year 1997 reprogramming to space Station is a set back to the plan."

I am wondering if, first of all, is that still a valid statement? I assume that it is, that it is a setback to the plan as you reprogram money from the Space Shuttle to the space Station. And it also flies a little bit in the face of your testimony earlier—and I understand what your testimony is here today. But doesn't the planning document itself, sir, illustrate that further reprogramming away from Space Shuttle safety is unwise and that it would be a setback to the plan of keeping the Shuttle safe?

Mr. TRAFTON. Let me say that the POP input is a program manager's recommendation to me as to what he or she would like to do the next year—actually 2 years, they are 2 years in advance. So it is my job and the job of my staff here—and we put the human space flight budget together based on POP inputs from across four centers and all of the programs that report to me. So everybody would like to do more. And we took that input and we said that what we're going to do is go forward with $100 million for phase two upgrades in the odd years. It is basically $75 million in Fiscal Year 1998 with $20 million for phase three studies.

The answer to your question, "Could they have done more upgrades with $190 million?" The answer is, "Yes, they could have done more upgrades." But I have to weigh things here, and I also have to ensure that the space Station stays on schedule. So we have to do what we think are the mandatory upgrades for as you've heard, obsolescence and safety for the Shuttle, and I've got to keep this other program on schedule. These two programs are joined at the hip now. You cannot talk about one without talking about the other.

So, I hope this answers your question. We limited them to $100 million in the out years, but we know that we have not impacted safety, and that has been confirmed by some outside eyes here.

Mr. NETHERCUTT. That is my concern, as I stated to you earlier. And I look at your testimony and it says that you don't plan to reprogram. Then I look at the POP plan and it says than any reprogramming, additional reprogramming would be a setback to the plan assuming that upgrades are a very important component of both the Shuttle and the space Station.

So I'm just hoping that there isn't some—I think that this Committee wants to be assured that there isn't some shortsightedness with regard to safety or erring on the side of space Station assistance at the expense of Space Shuttle safety. I think that is all important, and certainly Mr. Johnstone testified about the safety and the necessity of having the Shuttle program remaining very viable safety-wise.

Mr. JOHNSTONE. If I might add a statement that maybe would clarify a little bit. As of right now, I know, and I don't think our panel knows, of any safety implication upgrade that has been curtailed, deferred, not studied because of a lack of funds. Nothing has been held back to my knowledge. That's why we can say that everything is going ahead the way it should be.

Mr. NETHERCUTT. So you don't—you would disagree sir then essentially with the POP program recommendation that a reprogramming would be a setback to the plan. Am I correct in that?

Mr. JOHNSTONE. I don't know how they would reprogram. They are all funded and approved and they are going ahead. I'm not sure that I understand the POP statement except that, as Mr. Trafton says, it is just somebody's request for something that they might like to do in the future. Now whether those had safety implications.

Not all upgrades have safety implications. But I think that every safety upgrade has been funded, and they are mostly in that phase one group which are funded.

Mr. NETHERCUTT. I see. Thank you.

Chairman ROHRABACHER. Mr. Cook?

Mr. COOK. Thank you, Mr. Chairman, and I certainly want to thank the panelists for being here with us this morning.

I wanted to focus in on part of the testimony of Mr. Li in terms of some of the reprogramming of some of the research money. There was some indication of an $8 million reprogramming—or in that neighborhood—particularly relating to a potential long-term changes such as the liquid flyback boosters. And also, I think that you indicated in your testimony that potential funding requirements for some of these changes could be very, very substantial, $5-7 billion. Most of that related to the development of the liquid flyback booster. Is that accurate?

Mr. LI. That's correct.

Mr. COOK. I guess what I'm trying to understand in this research reprogramming, to what extent is this a realistic use of current type research dollars when the overall funding levels for NASA, if you look at Fiscal Year 1998 versus Fiscal Year 1999 there is about an $800-900 million decrease.

Is it realistic or is it even prudent to be spending research dollars on a cost—on a program that is as costly as this. And I'm wondering, as you respond to this, what implications—or what does this mean, I should say. What does this mean in terms of your belief in the viability of the X33 program and the overall life of the Shuttle?

Mr. LI. That's a heavy and big question, Mr. Cook. But I will try to answer that.

I think that as the Chairman has indicated, there are a lot of alternatives to extending the life of the Shuttle. One of the biggest components of extending the life of the Shuttle to 2030 assumes the use of the liquid flyback booster.

Now, I think that for a program as large as the liquid flyback booster, you would expect to have a very intense debate, obviously in this Committee, in which you would try to look at the alternatives to liquid flyback boosters. I think that the amount of money that they are using to study it is appropriate because it is going to provide the basis for that particular discussion. So, I think that the aspect of using carryovers to do studies is appropriate—if it is associated with the overall problem, which I think that this is, because it is a big issue of how long should this Shuttle be used for requires that particular input. So I do think that from the standpoint of the expensive nature of liquid flyback boosters, this is an investment toward that debate.

Now, that much said, I think that the use of carryovers should be limited to things that we can quantify and it should not be the seed for a lot of additional expenses. If this $4 million for the study of liquid flyback boosters becomes a larger amount that is associated with the liquid flyback boosters, then I can understand your concern. That should not be because that was not something that was discussed with the Congress in terms of their use of the money.

Mr. COOK. Well, I think that I am concerned about this current research spending without realizing really where we are going with it. And I still feel a sense of confusion because—especially those who feel like I do that the X33 is a great opportunity, and if there is something about that program that makes you feel that we can't really expect what we're already being told to actually happen with respect to the X33 for the overall lifetime of the Shuttle. I hope that you will really be very forthcoming with me.

And I've got to say this. As long as we're talking about studies of upgrades on the basic propulsion systems, why isn't there more of a discussion of improvement to the solid rocket motors that we are now using in the Shuttle? In terms of upgrading, I understand that there are several very key development plans that are being proposed in that regard. I'm just wondering, are you committed to moving away from the solid system?

Mr. LI. No. Obviously, NASA should be the one to address that.

Mr. COOK. And there are representatives from NASA here and please feel free to respond.

Mr. LI. My perspective on why they have decided to propose a liquid flyback as opposed to improvements to the solid is based upon the throttle back capability of the liquid that obviously—and we all remember the tragedy of Challenger—that if there is an unfortunate accident that should occur, a liquid flyback booster would perhaps provide additional risk mitigation that you could shut it off as opposed to a solid. I understand that there has been a proposal made by Thiokol to look at solids again, but we have not, at GAO, looked at that particular proposal.

Mr. COOK. And—I don't want to take more time if I'm already out, but if anyone from NASA could give us a comment, I would appreciate it.

Mr. TRAFTON. Sir, a couple of points.

We have formed a Space Transportation Council at NASA to look at our overarching investment strategy for the future of space transportation, and this is CODE M Space Flight, and CODE R the Aero and Space Technology folks together, one team, co-chaired. And this group reports directly to Dan Goldin, the Administrator. And they will decide where we go. And they will receive the results of these liquid flyback booster studies. And in fact, any upgrade money that I have in terms of studies for phase three and four, I'm turning over. It will be spent as this Committee directs.

We have been talked to by Thiokol. They have come in with a presentation. So, we have not decided that we are proceeding with liquid flyback boosters at all. All we're doing is a study to see were we to go forward, what would the requirements be? What would it look like? We've got Boeing and Lockheed Martin both under contract. We've agreed that we will not spend any more dollars in these studies without coming to this Committee and discussing such expenditures.

Mr. COOK. And just very quickly, if you were to put—if liquid flybacks looked like the greatest thing since sliced bread and butter, and you were to put them on all the Shuttle systems, aren't we talking about—what order of magnitude?

Mr. TRAFTON. We're talking probably in excess of $2 billion. And that would not be funded out of uncosted. We would come in and it would be a line item in the budget, sir. As will all upgrades beyond 2001.

Mr. COOK. Okay, thank you.

Chairman ROHRABACHER. Thank you, Mr. Cook. And we obviously have a vote on. Mr. Luther, and Ms. Jackson Lee, would you like us to reconvene after the vote?

Ms. JACKSON LEE. It's 10 minutes. I don't know if Mr. Luther has some quick questions. I can certainly finish before the—this is the only panel?

Chairman ROHRABACHER. Yes.

Ms. JACKSON LEE. I can certainly be quick, and Mr. Luther is before me, and I don't know if——

Mr. LUTHER. No, Mr. Chairman, I would be glad to yield, and certainly we can submit our questions in writing.

Chairman ROHRABACHER. Okay, thank you very much.

Ms. Jackson Lee—we have about 5 minutes at the most.

Ms. JACKSON LEE. Yes, and I certainly——

Mr. Chairman, let me thank you for this hearing. Obviously those of us in Texas in the Johnson Space Center area along with several of my other colleagues have a great interest in this area, and we appreciate your presence here. We had a very dynamic hearing if you will right before our good friend, Dr. Wolf, went off on his mission.

I think that it is extremely important that Congress does have an oversight responsibility. And my question to the gentleman, if you would—and I apologize if in your testimony you have already offered an explanation of sorts—but recognizing that there have been some problems in what we think are cushioning monies that are going to our Russian friends because it seems that they don't have the financial wherewithal, what comfort level do we have that other areas are not being deprived? In particular, the questions of our own safety processes are not being deprived because of funding sources being drained to support our Russian friends? And how much commitment that monies will be forthcoming? And if I can conclude with this last question, if someone can give me precisely the one, two, three, fours of the safety focus that you utilize in assuring us on the Space Shuttle safety issue.

So the first question is about are we draining ourselves by helping to support the Russians with our own dollars?

And with that, Mr. Chairman, I would ask unanimous consent to submit my opening statement for the record, please.

Chairman ROHRABACHER. That has already been done without objection.

And whereas the questions that Ms. Jackson Lee has posed goes to the heart of what this meeting has all been about, why don't we give each one of you 30 seconds with which to answer the questions that would be summarizing for her what we've said so far.

Mr. TRAFTON. I assure you, ma'am, that all Shuttle safety upgrades are being properly funded and that we are not taking monies away from the Shuttle program in a way that impacts the safety of the program.

Chairman ROHRABACHER. Mr. Johnstone?

Mr. JOHNSTONE. I duplicate or verify what Mr. Trafton has said. We find no evidence whatsoever that the funding transfer has impacted Shuttle safety, and I might also add that we believe very strongly that the Shuttle and the ISS safety are intertwined now. As Mr. Trafton said, "They are joined at the hip."

Ms. JACKSON LEE. Any slightest hint of impact and you would intervene?

Mr. JOHNSTONE. That's right.

Chairman ROHRABACHER. Mr. Li?

Mr. LI. Our position is also that safety has not been impaired. The projects that we looked at that were funded out of those particular programs had been proceeding. Many of them were complete or almost completed. At issue is something that a previous few minutes ago that Mr. Trafton talked, about that perhaps there might be some additional projects that would have been funded outside of the $190 million. However, the existing, on the planned projects related to safety, those were not impacted.

Ms. JACKSON LEE. Jim, it is good to see you and welcome.

Mr. ADAMSON. It is good to see you as well, Ms. Jackson Lee.

First to the fund transfer, relative to SFOC contract there has been no impact of any funding transfer to date. Our contract is unchanged. Our plans for safety enhancements and upgrades remains as it was before.

I would just like to add one other point. I think that people sometimes have a tendency to classify all upgrades as safety enhancements or obsolescence issues. And they are not. I think of several reasons to do upgrades. One for performance improvement. One for cost benefit of the operation which is one that the United Space Alliance is certainly interested in doing. And then of course there is the ongoing obsolescence and upgrades that any program has relative to obsolescence of vendors and supportability. So those are ongoing challenges that we will always be faced with.

To my knowledge, and I concur wholeheartedly with my friend Mr. Johnstone and Mr. Trafton, I know of no obsolescence or safety upgrade in the program that has been curtailed because of funds.

Chairman ROHRABACHER. We're down to about 3 minutes, and I want to thank Ms. Jackson Lee for her questions.

Ms. JACKSON LEE. Thank you.

Chairman ROHRABACHER. You will be able to ask further questions by writing.

And let me just say before we close that I would hope that the money—we have heard today that there is no impact on safety from these monies that are being transferred from the Shuttle. I would hope that the money that is being spent for upgrades is being spent for safety and upgrading capabilities of current Shuttle missions. And one thing that has concerned me—but we will be vigilant on this—if we are to have a Mars mission based on Shuttle technology, that has to be something that is based on a decision by the United States Congress not based on a decision by unelected officials. And let's come here and discuss it and let's make sure—if this indeed is going to be a purpose for the use of spending of our money, let's decide on that. But I better not find out that what's going on is that spending money to upgrade the Shuttle in order to hook us into a Mars mission without having any consensus by the elected officials.

So with that said, I want to thank each and everyone of you for your excellent testimony. This has been a terrific hearing and it was so because of your frankness and your completeness in your testimony, and we appreciate each and every one of you.

Ms. JACKSON LEE. Mr. Chairman, thank you for allowing me just one reason for being delayed as I was in a crime hearing talking about not legalizing marijuana. I obviously had to leave that for this was of higher authority.

Thank you very much, Mr. Chairman.

Chairman ROHRABACHER. Thank you. And this session is adjourned.

[Whereupon, at 11:30 a.m., the hearing was adjourned.]

[The following material was received for the record:]

NATIONAL AERONAUTICS AND SPACE ADMINISTRATION
OFFICE OF SPACE FLIGHT

SPACE SHUTTLE
FUNDING SUMMARY

	FY93*	DELTA	FY94*	DELTA	FY95	DELTA	FY96	DELTA	Estimated FY97	Estimated FY98
Uncosted Carryover (Beginning of Ye	266.0		388.4		435.9		328.2		467.5	399.4
New Obligation Authority	**3986.6**	*-11%*	**3554.4**	*-11%*	**3155.1**	*0%*	3143.8	*-6%*	2960.9**	2927.8
Available to Cost	4252.6		3942.8		3591.0		3472.0		3428.4	3327.2
Costs (Actuals/Est. Actuals)	3864.3	*-9%*	3506.9	*-7%*	3262.8	*-8%*	3004.5	*1%*	3029.0	3119.0***
Uncosted Carryover (End of Year)	388.4		435.9		328.2		467.5		399.4	208.2 ***
Uncosted in %	***10%***		***12%***		***10%***		***15%***		***13%***	*7%*

The Space Shuttle Program Mitigates NOA Perturbations through Prudent Cost Management

** - FY 1993 and FY 1994 funding normalized to current appropriation budget structure and includes June 1997 recission of $4.2 M of FY 1994 funds*
*** - Reflects Operating Plan transfer to Russian Program Assurance per NASA revised Operating Plan letter to Congress of May 5, 1997*
**** - FY 1998 Cost Estimate and resultant carryover reflects current budget planning*